Statistics for the Rest of Us

- Mastering the Art of Understanding Data Without Math Skills

By Albert Rutherford

www.albertrutherford.com

Copyright © 2023 by Albert Rutherford. All rights reserved.

No part of this publication may be reproduced, stored in a retrieval system, or transmitted in any form or by any means, electronic, mechanical, photocopying, recording, scanning or otherwise, except as permitted under Section 107 or 108 of the 1976 United States Copyright Act, without the prior written permission of the author.

Limit of Liability/ Disclaimer of Warranty: The author makes no representations or warranties regarding the accuracy or completeness of the contents of this work and specifically disclaims all warranties, including without limitation warranties of fitness for a particular purpose. No warranty may be created or extended by sales or promotional materials. The advice and recipes contained herein may not be suitable for everyone. This work is sold with the understanding that the author is not engaged in rendering medical, legal or other professional advice or services. If professional assistance is required, the services of a competent professional person should be sought. The author shall not be liable for damages arising herefrom. The fact that an individual, organization of website is referred to in this work as a citation and/or potential source of further information does not mean that the author endorses the information the individual, organization to the website may provide or recommendations they/it may make. Further, readers should be aware that Internet websites listed in this work might have changed or disappeared between when this work was written and when it is read.

For general information on the products and services or to obtain technical support, please contact the author.

ISBN: 9798391345831

First Print in the United States of America in 2023.

I have a gift for you...

Thank you for choosing my book, Statistics for the Rest of Us! I would like to show my appreciation for the trust you gave me by giving The Art of Asking Powerful Questions – in the World of Systems to you!

In this booklet you will learn:
-what bounded rationality is,
-how to distinguish event- and behavior-level analysis,
-how to find optimal leverage points,
-and how to ask powerful questions using a systems thinking perspective.

Visit www.albertrutherford.com for your FREE GIFT: The Art of Asking Powerful Questions in the World of Systems

Table of Contents

Chapter 1: Why You Need This Book   11

Chapter 2: Fundamentals Of Statistical Analysis   21

Chapter 3: Gathering And Interpreting Data   33

Chapter 4: P Values And Bayes' Theorem   45

Chapter 5: Statistical Thinking   53

Chapter 6: Applied Statistics In Real Life   65

Chapter 7: Visual Displays: Telling Stories Through Images   77

Chapter 8: Misinterpretation Of Statistics: 5 Common Pitfalls   93

Chapter 9: Data Manipulation And The Power Of Graphs	105
Chapter 10: Conclusion	113
Before You Go…	119
About The Author	121
Resources	123
Endnotes	135

CHAPTER 1: WHY YOU NEED THIS BOOK

Have you ever been watching tv and seen an advertisement for a product that convinced you to try it? The ad might have featured some statistics that made the product seem even more enticing (98% of people who tried this reported improved symptoms! On average, people lost 25 pounds on this medication! 87% of people interviewed responded favorably to our question!). Advertisers, politicians, and anyone else looking to make a point often cite numbers in their arguments, knowing that those numbers can sway the average consumer or voter. Most of us see a number that sounds convincing and believe it without question – 98% of people felt better on that? I've got to try it!

Numbers are a powerful weapon. The advertisers, politicians, and others who use numbers usually (but not always!) know they can be misleading. Maybe the ad citing 98% didn't tell you that the population they sampled included only people whose condition was in remission. Or maybe the "average" person on a weight loss drug lost 25 pounds, but they sampled one person who lost 150 pounds and five people who lost nothing. Statistics can be manipulated to present all sorts of arguments, only some of which are valid.

Sometimes misinterpreted or misapplied statistics can take hold and become a common belief. In the 1980s and 1990s, a study published in prestigious scientific journals concluded that left-handed people die, on average, nine years younger than right-handed people.[i] If you're a lefty, someone has probably quoted this statistic to you ominously. Reporters popularized some theories you may have heard, including those cited in the Los Angeles Times. "The researchers attributed much of the dramatic increase in accidental deaths to the fact that most machines are designed for right-handers. Certain neurological and immunological defects often associated with left-handedness were also thought to play a role in the shortened life spans."[ii] If you're left-handed, you're more likely to die in an accident, and there may even be something wrong with your brain!

But it turns out all those theories are irrelevant because there was a flaw in the study.

The researchers, Diane Halpern and Stanley Coren, noted that the proportion of lefties in the general population skews younger (a greater percentage of young people are lefties than older people). The LA Times reported, "In the population at large, about 9% of women and 13% of men are left-handed, but previous studies have shown a peculiar age distribution. At the age of 10, 15% of the population is left-handed. At 20, 13%. By 50, the figure drops to 5%. And by 80, it is less than 1%. 'That's what initially got us interested in studying this,' Halpern said."

Halpern and Coren drew an incorrect conclusion, though. Their research involved surveying the families of people who had recently died and asking if they had been left- or right-handed. When the results came back that a higher percentage of the younger people who had died had been left-handed, they concluded that something about left-handedness was to blame. But they failed to consider that many of the older people who had recently died probably would have been left-handed if they were raised today. For several generations, left-handedness was stigmatized, and school-aged kids who presented as left-handed were trained to become right-handed.[iii] Surveying the families of those who died didn't

13

yield accurate data about who was born left-handed and who wasn't.

In the left-handedness study, there was a hidden bias in the sample (fewer older lefties existed because of a reason that wasn't taken into account in the study). Here's another example of hidden bias affecting the outcome of a study: Imagine you go to Chicago and ask hundreds of people if they're fans of the White Sox. Keep in mind that Chicago has two baseball teams, the Cubs and the White Sox, and normally the Cubs attract many more fans than the Sox. In an average year, a typical Chicago baseball fan might say, "No, I don't root for the Sox." But if you ask the question at a time when the Sox are in the World Series against the Philadelphia Phillies, of course, Chicagoans are likely to say, "Sure, I'm a White Sox fan."

Is the moral of this story that you shouldn't believe anything you read, even in scientific journals? No! The moral is that you, the consumer of information, need to learn to distinguish good science from junk science. If something doesn't seem right about a statistic you hear, it probably isn't. One psychology and medical education professor made this point about the lefties-die-young theory: "If this were true it would be the largest single predictor we had of life expectancy - it would be like smoking 120 cigarettes a day plus doing a number of other dangerous things simultaneously."[iv] In other words, it's not plausible, and we should have known right away that something was off about this study.

This book will give you the understanding and tools you need to discern good science from bad science, reliable results from unreliable results, and valid statistics from misapplied ones. Knowing some basic statistics can help you become a better consumer, able to tell a false claim from a real one. It can enable you to determine when a statistic represents something true or neutral and when it is being skewed to make a certain argument. You can become more politically savvy, more aware of manipulation by advertisers, and more able to draw conclusions that help you live a better, healthier life. Statistics can help you become a better citizen.

This may seem like a bold claim, but you'll see that it's true as you read this book. This book will provide you with just enough knowledge to spot a valid claim from a misleading one, question when a statistic seems too good to be true, and argue back to people who try to use statistics to convince you of something you know can't be true. Here's a brief overview of what you'll learn in the coming chapters:

- Fundamentals of Statistical Analysis
- Gathering and Interpreting Data
- P-Values and Bayes Theorem
- Applied statistics in real life
- Statistical Thinking
- Visual Displays: telling stories through numbers
- Misinterpretation of statistics: the five pitfalls of statistics and how to avoid them
- Data manipulation and the power of graphs

This book comes at a time when data literacy is more important than ever. But what exactly is data literacy? What are the skills you will learn in this book? Wikipedia defines data literacy as "the ability to read, understand, create, and communicate data as information."[v] Kevin Hanegan of the Data Literacy Project thinks the definition needs to be more comprehensive than that, including the mindset to analyze and interpret data. He, therefore, defines data literacy as "the combination of skills and mindsets that allows individuals to find insights and meaning within their data to enable effective, data-informed decision-making."[vi]

We can define data literacy as both the skills and mindset to find meaning within data and the ability to understand and communicate data effectively. We will look at the process of gathering and sorting through data, the various measures and visual displays that communicate data, and how statistics influence your daily life. We'll also spend time analyzing how statistics can be misused to try to convince you of something and how you can be prepared to spot data falsifications.

According to a 2022 article in Forbes on why data literacy is so critical, 82% of 2,000 business leaders surveyed "expect all employees to have basic data literacy."[vii] If you're on the hunt for a job, you may be asked what you know about collecting or analyzing data. Beyond that, though, data literacy is critical to being an involved citizen of a democracy, where our beliefs and opinions shape our government and culture. In a world where we consume information nearly nonstop through our smartphones, we need to be literate about the information we are consuming.

Not only do we consume data nearly nonstop, but data about us is constantly collected and analyzed, whether we know and approve of it or not. Remember the Cambridge Analytica scandal in 2018, when we learned that Facebook had sold user data to Cambridge Analytica? The information that tech companies have about you is why you get targeted ads on social media sites, your smartphone knows what news you might want to see, and your music streaming service knows how to pick just the right song. Long before the technological revolution, data about listeners was the reason a radio disc jockey knew what to play next!

Here's a description from the Brookings Institute of how social media companies use our data and the potential consequences: "Social media algorithms...are engineered to provide users with content they are most likely to engage with. These algorithms leverage the large-scale data collection of users' online activity, including their browsing activity, purchasing history, location data and more." This allows for targeted content, which "allows the spread and cementing of misinformation."[viii] Many argue that our country has become so politically polarized because of this use of data to spread misinformation.

Whether or not you're aware, data surrounds us and is constantly being created about us. It's in our best interest to harness the power of data and learn what statistics can do for us.

On a less serious note, politically (but more serious financially), data literacy can help your company save money, as it did the Seattle Seahawks. That same Forbes article cites Mark Nelson, President and CEO of Tableau, a data visualization company (you'll learn more about data visualization in chapters seven and nine), telling his favorite story about how understanding data can help a company save money. According to Nelson, the Seattle Seahawks had received many complaints from fans about the audio quality in their stadium when data saved them:

> The Seahawks were about to start a multimillion-dollar renovation of their sound system to resolve the issue — but first, they dug a little deeper into the data. By examining customer complaints on a heat map, they discovered that only fans seated in the stadium's four corners were complaining about sound quality.
>
> The Seahawks subsequently discovered a flaw in the stadium's original design

that affected audio quality in those corners.

So instead of a multimillion-dollar overhaul of the entire sound system of the stadium, they just put extra speakers into the four corners. Sure enough, the fan experience surveys all went up.[ix]

So, before you go and build a brand-new stadium for your football team, gather some data and see what it tells you. The ability to analyze statistics might save you millions.

CHAPTER 2: FUNDAMENTALS OF STATISTICAL ANALYSIS

First, you might wonder what statistics is and why it is so complicated. Dictionary.com describes it as the science that "imposes order and regularity on aggregates of more or less disparate elements."[x] In layman's terms, statistics help us understand life and the world around us. It looks at large sets of numbers or populations and tries to make meaning out of them.

Merriam-Webster's definition of statistics gives us more to break down. According to them, statistics is "a branch of mathematics dealing with the collection, analysis, interpretation and presentation of masses of numerical data."[xi] Slow down and read that sentence again – there is a lot of information in it! Statistics involves *collecting*, *analyzing*, *interpreting*, and *presenting* data. Each of those steps can be done well so the results are true and valid, or they can be done sloppily or with bias, so the results can't be relied upon. You've probably heard about the results of scientific studies being rescinded or powerful people and companies taking back previous claims. That's most likely because they found a flaw in their methodology, often in the first step of *collecting* data. On the other end, advertisers and politicians are experts in *presenting* data so that it tells the story they want people to hear – not necessarily so that it reflects reality.

Statistics are divided into two large categories, each of which is divided into smaller categories (see the chart below). The two main branches of statistics are theoretical statistics, which deals with the theory and mathematics of data collection and applied statistics, which consists of using statistics to help us understand life. In this book, we'll be looking at only applied statistics.

Applied statistics is broken down into several categories, including **descriptive statistics**, which *describes* phenomena, and **inferential statistics**, which uses samples and probability to make *inferences* about populations. Descriptive statistics include terms you probably learned about in upper elementary or middle school, including mean, median, mode, and range. Inferential statistics look at populations and samples. Here's a simple diagram of these two main branches of applied statistics to help you keep track:

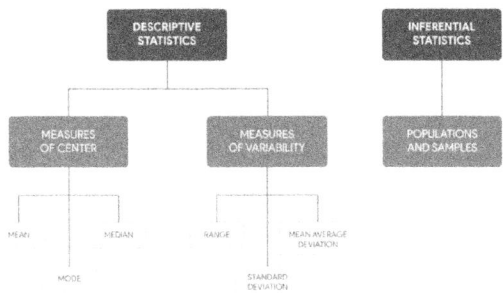

Before we delve deeper, you'll need to know a bit more about these basic terms that we will examine in later chapters. Here are some definitions you can reference, flipping back to this list anytime you need to remind yourself what a term means.

DESCRIPTIVE STATISTICS

Descriptive statistics give us a snapshot of what the data is. We can draw inferences from descriptive statistics, but they are meant simply to describe and summarize large data sets in numbers that are easier to manage. Descriptive statistics can be used on both

quantitative and **qualitative** data. Quantitative data involves quantities – numbers, and measures. Qualitative data involves attributes or phenomena that are measurable but not quantifiable. In other words, they can be described or labeled but not with numbers. An example of quantitative data is the ages of people in an office or the amount (in dollars) of debt people hold – this information can be represented with numbers. Qualitative data would be, for example, the favorite food of everyone in an office or the types of cars purchased last year. Qualitative data is a collection of categories, not numbers.

Both of these types of data can be described with descriptive statistics. Imagine you gather data from households in one particular neighborhood, for example. You might gather household income, which would be quantitative data, and also the color of the car(s) owned by each household, which would be qualitative data.

The following terms are most often used with quantitative data. These terms help us make sense of large data sets that would otherwise be overwhelming to sort through.

Measures of center (or measures of central tendency): calculations that lend information about the average or center of a data set.

Mean: the average of a set of two or more numbers. When you hear "average," this is almost always what people are referring to. This is found by adding all the numbers and dividing the sum by the number of terms in your set. You can think of it as distributing the data equally; if everyone or everything that was measured had the same amount of whatever you were measuring, what would that amount be?

Here's a simple example: There are five kids with cookies. Four kids have 1 cookie each, and the fifth has 9 cookies. If you redistributed the cookies so that each kid had the same number, they would each end up with 3, which is the mean number of cookies. We can find this by adding up all the cookies, then dividing by the number of terms (kids) in our set:

$$\frac{1+1+1+1+9}{5} = 3$$

The mean number of cookies the kids have is three. This doesn't mean that any one child necessarily has three cookies, but rather that if we distributed the cookies equally among all the children, each of them would have three.

Median: the middle point in your data set if you list the terms in order from smallest to largest (or vice versa).

Example: There are 5 people in a room, ages 6, 29, 17, 63, and 2. Arrange the numbers from least to greatest to find the middle term:

2, 6, **17**, 29, 63

The median age in this set is 17. Notice that you could arrange the numbers from greatest to least (63, 29, 17, 6, 2) and still have a median of 17.

If you have an even number of terms, you will end up with two numbers in the middle; to find the median, find the mean (average) of those two middle numbers.

Example: A 6th person joins the room, so now the people in the room are ages 6, 29,

17, 63, 2, and 41. Arrange the numbers in order again, then locate the middle two terms:

2, 6, **17, 29**, 41, 63

The mean of the two middle terms (17 and 29) is 23 because $\frac{17 + 29}{2} = 23$. The median of this set of numbers is 23. Notice that nobody in the group is actually 23; just like with the mean, it doesn't matter if the median does not represent an actual person or data point.

Mode: the number that appears most often in your set. There can be more than one mode if two or more numbers appear the same number of times.

Example: A survey asked ten families how many pets they owned. The answers came back: 1, 1, 3, 2, 2, 5, 1, 1, 5, 2. The mode would be 1 because that number appears more often than any other number in the set.

The mode can give you an idea of how common something is. None of these measures alone will tell you everything you need to know about a data set. However, when all three are used together, we get a better descriptive picture of a large data set.

Measures of variability: calculations that lend information about how varied the data points are

Range: the difference between the largest and smallest numbers in the set. You can find this by subtracting the smallest number from the largest number.

Example: In the group of six people mentioned above, whose ages were 6, 29, 17, 63, 2, and 41, the range of ages is 61 years, since 63-2=61.

The range tells us how wide the spread of data points is. Again, this measure is most useful in conjunction with other measures. If we were told that the median age in a group of six people was twenty-three, we would have no idea what the actual ages in the group were. The people surveyed could have all been in their twenties or could have ranged from zero to one hundred years old; both would yield the same median. The range tells us how varied our data points are.

Other measures of variability are **standard deviation**, **interquartile range** (also known as IQR), and **variance**. These are all important measures you would learn about in a statistics course and need in a sophisticated study. For the purposes of this book, however, it is enough to know that these measures of variability exist and lend important insight into the data.

Two additional terms that might come up in looking at variability are clusters and outliers. **Outliers** are data points that stand outside the main cluster of data. For example, if you survey people about how big their houses or apartments are, most people's homes will fall within a certain range, but a tiny house dweller or an owner of a mansion might give you data points that are outliers.

Outliers are notorious for skewing measures of the center, particularly the mean. This is why the median is more often used when reporting income. For example – those in the top 1% of the country's earners make so much more than the rest of the population that their incomes pull the mean higher. Outliers on the low side can also pull the mean down. If, for example, there was a group of thirty-year-olds and one baby at a party, the mean age of the group might be around fifteen, which misleads us about the age of the partygoers. In this instance – and in most instances where there are significant outliers – other measures of the center (like the median) are more useful.

Clusters are exactly what they sound like: where notable groups of data fall. If you surveyed every person in the United States about how many pets they owned, we could expect clusters of data (answers from different people) around the numbers 0, 1, and 2. Not as many people have more than two pets as do people that have two or fewer. Clusters and outliers are often used together when looking at the **shape** of the data. The shape literally means what shape, or what kind of curve (if any), you would see if you plotted all the data points on a coordinate grid. The most well-known data shape is the bell curve, where most data points fall somewhere in the middle of the range, with data points tapering off at the high and low ends of the range.

INFERENTIAL STATISTICS

Inferential statistics help us make sense of large populations that would otherwise be too difficult to survey or study. In inferential statistics, a study is done using a smaller representative sample of the population, and then inferences about the results are made to draw conclusions about the larger population.

Population: Exactly what you think it means – every single member of a group. If you have a statistical question about people between the ages of twenty and forty in the United States, the population is every single person between twenty and forty in the U.S. If you're studying a class or a family, populations can be small and manageable, but often they are very large, which makes surveying them unrealistic.

Sample: a representative subset of a population. Samples are used in studies because they are much more manageable than entire populations. The tricky part of sampling is that the sample *must be representative* of the entire population, meaning it must share the same characteristics of it in the same proportions. If you want to find out what percentage of the U.S. population voted for a Republican candidate in the last election, for example, it wouldn't be reasonable to survey every single voting resident of the U.S. Instead, you would select a sample that has about the same percentage of registered Republicans and registered Democrats as the general population does. If you only polled Republicans, your results wouldn't be valid. The subgroups in your sample have to be proportional to those subgroups in the larger population. We'll look at this in more depth in future chapters.

Now that you have a beginning understanding of some of the main terms in statistics and why it is so important for you to understand statistics, let's take a look at the process of gathering and interpreting data.

CHAPTER 3: GATHERING AND INTERPRETING DATA

Just as scientific experiments always start with a hypothesis, data analysis always starts with a "why." Before beginning to gather data, you need to have a reason for gathering the data. Are you trying to figure out a trend in a certain population? Determining preferences for something to help you make a decision or advertise a product? Wondering what you can learn by looking at averages from a certain topic?

The question you have will determine what process you use to gather data, and this will, in turn, determine how clean and accurate your data is. Each step of the process is critically important to get valid results. Statistical analysis can be thought of as a five-step process: Define the question and figure out methodology; collect data; clean and summarize the data using descriptive statistics; process the data and apply hypotheses; make inferences and apply findings.[xii] Here's a visual to help illustrate this process:

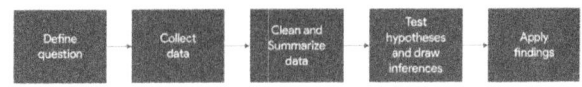

Step 1: Define the Question

Once you have figured out what you are trying to find out, you need to spend time crafting a question. Is your question quantitative or qualitative? If it is qualitative, how can you word it to get clean, unbiased results? This is a crucial step in the process. Questions can be written to lead people towards certain answers, in which case you won't get a neutral, unbiased result. Consider the following questions:

"On a scale of 1 to 5, how great did you think the speech was last night?"

"Which of our brand's products is your favorite?"

"Do you support proposition A to make our city government more fiscally responsible and decrease the massive debt it has accrued?"

All three of these questions assume something or lead you to a certain answer. What if, for the second question, you didn't like any of the brand's products, but you had to pick one in your response? This is a loaded question, assuming that you like all of the brand's products. The third question is a leading question, meaning the phrasing leads you to select one answer over another. The question is phrased in such a way as to lead you to say yes. If you have someone in your life who guilt-trips you ("Do you really want me, your elderly mother, to drive six hours to see my grandchildren, instead of you hopping on a plane to come to visit me???"), you're probably aware of this type of question!

Questions also need to be crafted so that you get clean and clear data. Imagine you want to find out how many people in a sample set enjoy potato chips. If you asked, "Do you usually order chips when you order a sandwich?" you wouldn't know if your respondents actually liked chips or if they just liked getting chips with their sandwich. Asking a more direct question without complicating information would get you a truer result.

Step 2: Collect Data

Once you have figured out what you're curious about and crafted a question, you're ready to collect data…But wait! Before you can actually collect the data, defining a sample is the next critical step. The sample you select needs to have the characteristics in the same proportions that the full population has. Watch out for hidden biases in the sample! The left-handed study, for example, had a hidden bias in the sample, leading to inaccurate results, inaccurate conclusions, and possibly thousands of lefties who feared they would die young simply because of their left-handedness!

One of the most effective ways to get a representative sample is through a process of random sampling. Random sampling means that the sample group is chosen completely randomly (often by computer); every person in the population has an equal chance of being selected as part of the sample group. If the population is relatively homogeneous, random sampling might work well.

Imagine, for example, that you want to find out what percentage of teenagers in the

United States chew gum. It's probably safe to say that gum is universally available in the U.S.; it's not particularly expensive, and you can find it at any grocery or convenience store. If you polled a random sample of teenagers, they should all have about the same access to gum as anyone in the wider population does.

On the other hand, imagine you want to find out what percentage of teenagers have traveled outside of the country. Determining a sample for this study would be trickier, as not every teenager in the U.S. has equal opportunities to travel outside the country. Some teens who live right near the borders of Mexico and Canada may be more likely to have traveled outside of the country, as would teens from wealthier families; teens without financial resources who live far from the borders are less likely to have left the country. A random sample, then, won't necessarily be representative of the general population. It *could* be, but there would be no way to guarantee that the sample wouldn't include (by chance) mostly people who live in border towns, for example.

Stratified random sampling is a more precise way of sampling when you have a heterogeneous population. Unlike truly random sampling, stratified random sampling accounts for the subgroups in a population.[xiii] Each subgroup is a *strata*, or a layer of the population distinct from the other layers. In the imagined study of teenagers traveling outside of the country, you would need to define the strata that impact whether or not they have experienced this. Maybe you would want to define the strata by economic status, maybe by geographic location, or something else. You can see why defining a sample can be such a tricky task! You would then use a proportional number from each strata in your sample.

Here's a simpler example for understanding stratified random sampling. Let's say you want to find out what percentage of the population of a certain city watches basketball on television. You think gender might be a factor in this, so it is important that your sample have the same proportion of males to females that the full city population does. For example, if the city was 55% male and 45% female, your sample would also need to be 55% male and 45% female. If there are non-binary members of the population, that might comprise a third strata that also needs to be represented proportionally.

Here's a visual representation to help explain stratified random sampling[xiv]:

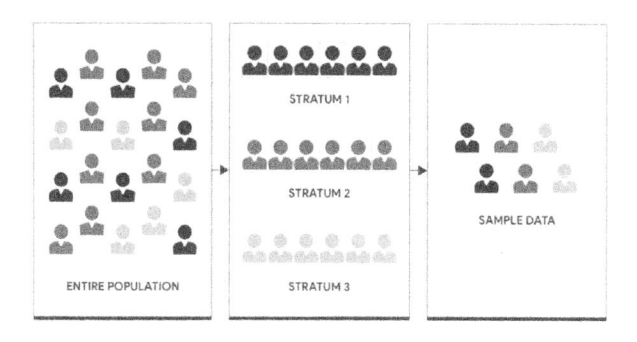

If you conducted a completely random sample of the population shown here, you might end up with all gray people (stratum 2) or light gray and dark gray (strata 3 and 1) but no gray, for example. Determining the strata and then selecting a proportional number from each of them means that your sample will have the same proportions of each strata as the entire population. In this example, there are an equal number of dark gray, gray, and light gray people, so the sample also has an equal number of them (2 of each), making it proportional to the larger population.

Other types of sampling draw on probability but offer slightly more control than straightforward random sampling. For our purposes, however, understanding the difference between purely random sampling and stratified random sampling can help us understand how important this first step in data collection is.

Some types of research call for non-random samples – say, for example, a drug company is testing an effect of a new medication on a targeted group. These are usually more exploratory or investigative studies rather than studies that look for trends in large populations.[xv] This is why figuring out your question and the type of data you need is critical. The question determines the data collection method that will be most useful and unbiased.

Now that you have your question defined and your sample identified, you are ready to go out and collect the data. If you have participated in a marketing study or answered a survey about something, this is the data collection step. If you have been careful up to this point, data collection should go smoothly, although you will still need to "clean" your data in the next step.

Step 3: Clean and Summarize Data

After you have collected your data, you're ready to analyze it, right? Not so fast. First, you need to look through your data and make sure it is clean. Cleaning data is exactly what it sounds like: looking through it to make sure it doesn't have any errors or duplicates, that it was collected accurately, and that it's in the right format for your computer (or you) to analyze.[xvi]

With small data sets, cleaning it might be as easy as looking it over to ensure there are no mistakes or missing pieces of information. With larger sets, though, cleaning can be a lengthy process. Particularly with data that a computer or automated process has collected, you're likely to have duplicates or missing values.

Imagine you survey a sample of a population about a certain topic using Google Forms as your survey tool. Looking over the results, you might find that some people forgot to enter their last name or typed their address where they were supposed to type their age. Maybe one person filled out the survey twice. People make mistakes, so it is inevitable that the data will need some cleaning.

The other critical part of the cleaning process, especially if you will be using a computer program to analyze the data, is making sure that it is all in the correct format. If you try to run an Excel program to find the mean of a set of data, and one entry is in the wrong cell (like an address where age should be), your results will be inaccurate or unreadable. Data cleaning can be tedious, but it is critical to ensure your analysis is valid.

Now that your data is clean, you're ready to analyze or "run the numbers." This is the fun part! This first pass at analyzing can give you a glimpse at the answer to your question. Using descriptive statistics, you'll find all the numbers that help you make sense of the data you just collected. Since you won't be sure yet which of these measures will give you meaningful information, you should look at as many as you can. Excel and Google Sheets have formulas that can look at data in many ways, giving you all sorts of tools for analyzing it.

Step 4: Test Hypotheses and Draw Inferences

Now that you have the data summarized, you can look at those numbers and start to make sense of them. What are the numbers telling you? Try to look at them with an unbiased eye since they may not be telling you what you expected or hoped they would.

Here are some questions to ask as you look at your data:

- What are the mean, median, and mode of the data? What is the range or other measures of variability? Does one of these tell me more than the other? Are any of these misleading or not useful in this situation?
- Are there outliers or clusters? What do these mean or tell me? Are the outliers skewing the data? Should I try running the numbers again with the outliers removed, so I can see the clusters more clearly?
- Is the data bearing out my hypothesis? Or is it telling me something different? Do any of the measures have statistical significance, or are they all so minor that they don't really mean anything?
- Is there something completely unexpected in my data? If so, am I sure I cleaned the data properly and ran the right formulas? If I am sure, why might that unexpected thing be there?

Step 5: Apply findings

In this last step, you might take the findings from a sample and apply them, drawing inferences and conclusions about the larger population. Did the study of your sample group show that people with red hair prefer dark chocolate to milk chocolate? Then – assuming you did every previous step correctly – you might conclude that red-headed people in the general population prefer dark chocolate to milk chocolate.

If you hear the results of a study on the news, this is most likely the step you are hearing. At the point of reporting, the study has been completed, and those who conducted the study have drawn conclusions. This sounds straightforward, right? Simply take your findings and apply them on a larger scale. It is easy to misinterpret findings, though, and even easier to present findings in a misleading way. Chapters six, seven, and eight will look more at how statistical findings can be manipulated in visual displays, trying to tell stories that aren't borne out by the data.

CHAPTER 4: P VALUES AND BAYES' THEOREM

The title of this chapter may have you running, but don't worry! Yes, these are terms used in advanced statistical analysis, but we're going to look at how a basic understanding of them can help you in your everyday life. An understanding of p-values and Bayes' Theorem will help you become a better interpreter of statistics you come across and a more analytical thinker in noticing patterns of behavior in your own life.

The first thing to understand before looking at either p-values or Bayes' Theorem is probability. Don't worry; this won't be a rehashing of your 7th-grade math class. This will be a short probability lesson to ensure you get the basics. Probability is how likely something is to happen. It is expressed as a fraction, a decimal, or a percent. Probability is always between zero and one as a fraction or a decimal. Zero means something absolutely will not happen; one means it is guaranteed to happen. As a percent, probability is between zero and one hundred, with one hundred meaning absolute certainty.

We can find the probability of something happening by taking the number of "favorable outcomes" (the thing we're measuring or hoping to get) and dividing it by the number of total possible outcomes. We call the thing we're measuring an "event." For example, if we're looking at how likely we are to get heads when we flip a coin, the coin landing on heads is an event.

$$\frac{Favorable\ outcomes}{Total\ possible\ outcomes} = Probability\ of\ an\ event\ happening$$

In the case of a coin flip where we are hoping to get heads, favorable outcomes would be one (there is only one way to get heads when you flip a coin), and total possible outcomes would be two (you can get two possible results, heads or tails). The probability of the coin landing on heads would be ½, which we can also express as .5 or 50%.

Put simply, p-values tell us how likely it is that something happened for a reason, not just because of random chance. Anytime you hear the phrase "statistically significant," you hear an analysis of the p-value. Calculating p-values involves calculus and something called the Null Hypothesis but knowing that it gives us a sense of statistical significance is enough for our purposes. Another way of thinking about it is that p-values tell us how likely it is that another experiment would have the same results. If you looked at what ten people ate for lunch on Tuesday, for example, and saw that 80% of them ate tuna, that is probably a result of random chance (though, who knows, maybe people like to eat tuna on Tuesdays!). Without actually calculating it, we could guess that the p-value for this study would tell us that, if you ran the study again next Tuesday, you'd most likely not get the same results.

P-values are a probability and thus are expressed as a number between zero and one. *Lower* p-values mean that results are more statistically significant; higher p-values mean the results may be due to chance. Different fields and academic journals often have their own criteria for what p-value means something is statistically significant. Still, usually, the p-value needs to be less than .05 for something to be considered statistically significant.[xvii]

So what does this mean for you? First of all, if you hear about a study, always look for the statistical significance or p-value. A very low p-value, close to zero, tells you that you're safe to believe the results of the study. On the other end of the spectrum, you may hear the results of a long-awaited study in the news that says the results were not statistically significant. This means that the researchers aren't confident that their results mean anything important!

You can also use the idea of p-values informally in your own life. Before you assign causation to something, think about how likely it is that that thing happened by chance. Have you noticed, for example, that it rained the last five times you wore brown socks? Does this mean that wearing brown socks causes it to rain? No. The fact that it rained when you were wearing brown socks can be explained by random chance. If, on the other hand, you got a stomachache the last five times you ate broccoli, you might want to ask yourself how likely it is that you get a stomachache when you eat anything. If you don't normally get stomach aches – if they're not likely to happen by random chance – then it might be statistically significant that you've gotten one when you ate broccoli, and you might want to avoid broccoli from now on!

What we just looked at is called frequentist probability because it relies on measuring the frequency with which something occurs. Another way of analyzing likelihood is based on Bayes' Theorem, which gained momentum in the late twentieth century.[xviii] Bayes' Theorem looks at how likely something is to happen, *given that* something else has already happened. It is a way of measuring conditional probability and is often used in business models and market analysis.[xix] You can also use it in your everyday life for decision-making.

A classic example that draws on Bayes' Theorem is the Monty Hall problem. The problem, based on the game show *Let's Make a Deal* (originally hosted by Monty Hall), goes like this: You're a contestant on a game show. In front of you are three doors, and you're told that behind one of the doors is a new car, and behind the other two doors are goats. You get two guesses. For your first guess, you choose one door, and the host (who knows where the car is) opens a different door to reveal a goat. For your second guess, do you want to stick with your choice or switch to the other still-closed door?

Our intuition tells us that we are just as likely to get the car if we stick with our first choice as if we switch to the other door. Each of the three doors has a ⅓ chance of having a car behind it, right? But the host's action – opening one door to reveal a goat – now has to factor into our decision-making. This is an example of conditional probability. An event happened (one door was opened), and so we have a new set of information: the host knew that the car wasn't behind the door he or she opened. How likely is it that the car is behind the door we chose *given that* the host revealed it is not behind the opened door?

Believe it or not – and most people do not believe it – you are more likely to win the car if you switch your selection to the other unopened door. An analysis of this problem appeared in *Parade* magazine in 1990 and caused quite a stir, with thousands of people, including mathematician Paul Erdos, writing to the magazine to argue that the answer provided (switching doors gives you a greater chance of winning) was incorrect.[xx]

A Bayesian analysis, done most easily by computers, reveals that you have a ⅔ chance of winning the car if you switch your guess to the other door. This doesn't seem to make sense intuitively, but our and the host's actions need to factor into the decision. In other words, given that something else has already happened, the probability of something happening differs from what the probability was at the start of the challenge.

The formula for Bayes' Theorem is complicated for most laypeople, which is why Bayesian analysis didn't become popular until the computer age. In your everyday life, you don't need to know the formula or even have a solid mathematical understanding of the theorem to use its basic principle. You may, in fact, already make decisions loosely based on the logic behind the Theorem, taking into account events that have already happened and what they can tell you about what might happen.

Let's imagine a dating scenario. You meet someone you have a lot in common with and immediately like them. Without you being aware of it, your subconscious is calculating how likely it is that this relationship will succeed (in whatever way you define "succeeding" at that moment). On your fourth date, they reveal something about themselves – let's say they order an expensive bottle of wine. This event now influences how you think about them. In your subconscious, you are calculating how the expensive wine purchase factors into what you previously thought of them. You are calculating the probability that the relationship will succeed *given that* they have now purchased an expensive bottle of wine.

The result of this new conditional probability will be different for everyone and in every circumstance. Maybe you had bad experiences in the past with people who bought expensive bottles of wine, so history factors into your equation and you now see your relationship as less likely to succeed. Or maybe you associate buying wine with traits that you admire in a person, so the action makes the probability that your relationship will succeed more likely. This is a non-mathematical model, but it's one that illustrates how new information can change the likelihood of an event occurring.

This might seem obvious – that every action or event changes the likelihood of a certain outcome – but Bayes' Theorem represents a whole new statistical analysis method. It is particularly useful in the field of investing, as investors constantly incorporate new information and their existing hypotheses to predict what the market is likely to do.[xxi] For the average person, incorporating the fundamentals of Bayes' Theorem into your life means asking yourself how the new information you get changes things and trying to think mathematically or statistically about it. That's what we'll look at in the next chapter.

CHAPTER 5: STATISTICAL THINKING

We now have an idea of why it's so important to understand a bit about statistics. Despite knowing how valuable statistics can be, humans still tend to believe what they want to believe. This chapter will tell you more about why this happens and define what it means to think statistically. In the process, it will help train your brain to avoid the common traps of human psychology and instead look for the truth in what numbers can tell you.

If you've ever brought up something you saw on the news with an elderly relative, you've probably heard a sigh and a response like, "Crime is so bad these days. I just don't know what's wrong with people now!" Most people that you survey on the street will tell you that crime is up in their world. They may cite the news or stories from friends who live in nearby towns or cities. "Crime is so much worse than when I was young," they might tell you, or "Kids these days just don't know right from wrong." But this impression is almost universally incorrect. Statistical research shows that violent crime in the U.S. has been declining since 1993. Most people are just plain wrong. Here's a graph of what the Pew Research Center has found[xxii]:

VIOLENT VICTIMIZATIONS PER 1,000 AMERICANS AGE 12 AND OLDER

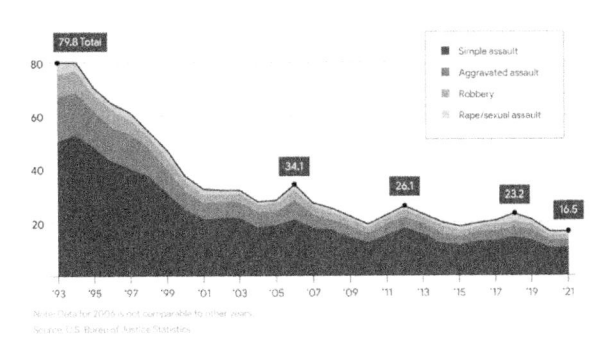

Despite some small spikes, violent crime has *been significantly* down since 1993. People still believe that crime is getting worse, though. According to the Pew Research Center, "In 22 of 26 Gallup surveys conducted since 1993, at least six-in-ten U.S. adults said there was more crime nationally than there was the year before, despite the general downward trend in the national violent crime rate during most of that period."[xxiii]

There are several reasons people might believe something that isn't true. One reason is the media. The United States (and, indeed, the world) has seen an explosion of misinformation on social media as well as on some news outlets. But an even simpler reason is that people tend to rely on intuition, which often leads them to be wrong. One effect psychologists study is the *availability heuristic*, which refers to how easy it is to draw up an example of what you believe.[xxiv] Most people can easily think of an example of a crime they heard about or witnessed – think of an elderly relative beginning a story with "I saw this horrible thing on the news the other day…" and ending it with "Crime is so bad these days!" The availability of an example convinces us that something is true. If I hear about crime happening, that must mean crime is prevalent in my community; therefore, crime is up! But this just isn't true. As much as we want to rely on our instincts, intuition often leads us to incorrect conclusions.

Human emotions can also lead us to illogical conclusions. If you or someone you know has a fear of flying, you know how irrationally our emotions can make us behave. You probably know by now that a person is far less likely to die in a plane crash than in a car crash (1 in 1 million times vs. 1 in 5,000 times, according to most sources). And yet there are plenty of people who won't get on planes but will happily drive somewhere.

Psychologists attribute this seemingly illogical behavior to a number of factors. Along with the availability heuristic, or how easily you can come up with an example of what you believe (for example, that the plane you get on is certain to crash), is the idea of control. When we drive, we feel in control of what is happening. We can try to drive more safely to lessen our chances of an accident. We have no control over the outcome of a flight, and therefore we tend to feel more anxious about it, which makes us behave less rationally.[xxv]

If you've ever tried to convince the aerophobic that they're being silly and should just get on that plane, you know it's not easy. Human beliefs and emotions are very powerful, and it takes a significant effort to see past one's anxieties, beliefs, and biases and believe the facts. So how can you make sure that you're not swayed by intuition, that your anxiety doesn't drive your decision-making, and that you don't get caught up in conspiracy theories that prey on your emotions?

Anytime you hear a statistic or see a chart, you can ask yourself several questions before you accept the information as presented. Each of these questions can be answered mathematically, but there are also ways that you, the consumer of information, can check to see how reliable and valid the information is. Keep these questions in mind when you encounter statistics:

Question 1: How was the data collected, and did the data collection method match what the study was trying to figure out?

This question gets at a study's validity, which means measuring what it is supposed to measure. Sometimes results are published, but sloppiness in the way the data was collected, the sampling, the interpretation, or another step in the process can lead to invalid results. Think back to the study of left-handed people's mortality in chapter one. The study's authors thought they were measuring the average age of mortality for left-handed versus right-handed people. However, the results didn't tell us that information because of how they identified their sample (looking at recent deaths) and conducted their survey (asking relatives if the people had been right- or left-handed).

Another popular debate for validity lies in standardized tests. Debates have raged for years over whether standardized tests are a valid measure of intelligence or academic achievement, with many arguing that biased test questions and outside factors invalidate certain test results. The SAT and ACT are two of the most hotly debated tests. Many colleges across the United States no longer require one of these tests for admission since many factors – like access to test preparation and private tutors- can influence a student's score.

While it is true that the SAT and ACT can have inequitable outcomes due to outside factors, several analyses have shown that SAT scores correlate with IQ scores and are a good predictor of college success.[xxvi] This doesn't necessarily mean colleges should be basing admission on these scores. Still, it shows us how hard it is to determine a test's validity, particularly when it has been entrenched in our culture for many years.

Question 2: Can the results be replicated in another study?

This question asks about a study's reliability or how likely you are to get the same result if you performed the test again. Let's imagine you want to study how popular *Star Wars* is in the general population across the

United States. You decide to ask a sample of people whether or not they like the *Star Wars* series. The answer you get from your sample – let's say 99% of people – is overwhelmingly "Yes!" So that means 99% of the population of the U.S. must love *Star Wars*, right?

It turns out that the sample you asked was comprised of people leaving a movie theater after the midnight opening of the latest *Star Wars* installment. These are super-fans – it's safe to say that most of them love *Star Wars*! We don't know from the study you performed if 99% of the public loves *Star Wars* or not. The sample you chose was not representative of the general population, and therefore the results of your study are unreliable. What you were measuring was what percentage of that particular group (midnight viewers of the latest *Star Wars* movie) loves *Star Wars*, not what percentage of the general population does. If you replicated your *Star Wars* study but instead sampled a pool of people leaving a coffee shop, you would probably get much different results, which would show that the study wasn't reliable.

A scientific study, a survey, or a test should be both reliable and valid. If it lacks one of these traits, its results can't be relied upon.

Question 3: Given the data, is the conclusion drawn a logical one?

We'll examine this question in greater depth in Chapter 8 when we look at the common pitfalls of statistical analysis. One-way statistics can be misleading is that they can show you correlation but not causation. In other words, just because two things happen to be linked doesn't mean one *causes* the other. Here's a fun example of correlation:

DIVORCE RATE IN MAINE
correlates with
PER CAPITA CONSUMPTION OF MARGARINE

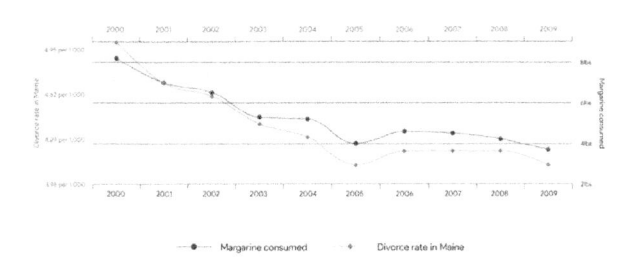

- Margarine consumed ◆ Divorce rate in Maine

Based on this data, you might be led to believe that the divorce rate in Maine is caused by eating margarine, or maybe vice versa. This graph shows a correlation, but drawing a conclusion about causation, in this case, is illogical.[xxvii] We'll get into other problems with this graph in a bit, but your first question to yourself needs to be if the results make sense.

Question 4: How is the data presented, and is the presentation misleading in any way?

This question gets a whole chapter to itself (Chapter 9) because misleading statistical presentations are, unfortunately, very common in the media. Advertisers, politicians, and anyone else trying to convince you of something make their point by exaggerating small differences. Sometimes they make their point by emphasizing differences that don't even exist!

In 2015, Representative Jason Chaffetz of Utah used the following graph in a Congressional hearing questioning the president of Planned Parenthood.[xxviii]:

PLANNED PARENTHOOD FEDERATION OF AMERICA
ABORTIONS UP - LIFE-SAVING PROCEDURES DOWN

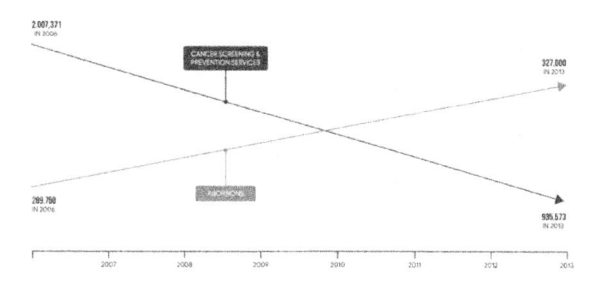

2,007,371 IN 2006

CANCER SCREENING & PREVENTION SERVICES

327,000 IN 2013

289,750 IN 2006

ABORTIONS

935,573 IN 2013

2007　2008　2009　2010　2011　2012　2013

The graph appears to show that, between 2006 and 2013, abortions provided by Planned Parenthood skyrocketed, surpassing the number of cancer screening and preventative services they provided. That is a powerful visual!

But if you look more closely at the graph, you'll notice there is no y-axis. The y-axis provides the scale of what is being measured; the fact that there isn't one tells us that the placement of these arrows doesn't necessarily mean anything – the visual display may not match the real numbers. Take a look also at the numbers at the end of each arrow. The pink arrow says 935,573 cancer screening and preventative services were provided in 2013, while the red arrow shows 327,000 abortions provided that same year. So why is the red arrow so much higher than the pink when the numbers show that nearly 600,000 more cancer screening and preventative services were given than abortions that year?

This was an example of a misleading presentation meant to convince you of something. Politifact made another graph using the data, this time with a defined y-axis that has a base of zero and the data points located accurately on the graph:

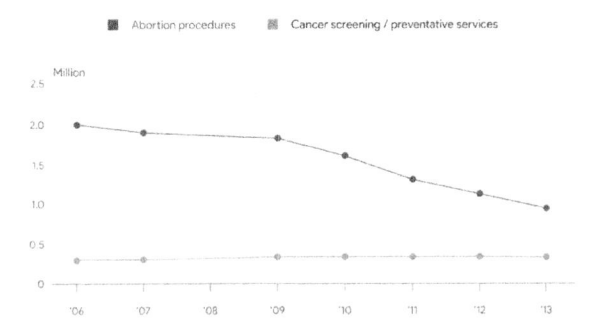

PLANNED PARENTHOOD SERVICES

■ Abortion procedures   ■ Cancer screening / preventative services

Cancer screenings and preventative services did, in fact, decline over that time span, and abortions rose slightly, but abortions came nowhere close to surpassing preventative services. Politifact reported what one communication professor thought of the original graph: "'That graphic is a damn lie,' said Alberto Cairo, who researches visual communication at the University of Miami. 'Regardless of whatever people think of this issue, this distortion is ethically wrong.'[xxix]

Graphs should always have a title and a clearly defined scale. If the graph is a coordinate plane with two axes, those axes should be labeled, and the scale should be consistent on each (it doesn't need to be the same on the two axes, but each axis needs to have a consistent scale). Take a look again at the graph showing a supposed correlation between divorce rates in Maine and margarine consumption. Notice that there are two scales for the y-axis, one on the left for the divorce rate and one on the right for pounds of margarine consumed. These two scales tell us that the quantities were measured in different ways. The only thing that is true in this graph is that the shape of the data – where the two phenomena rose a bit and where they fell – roughly matches. Be on the lookout for this kind of dual labeling, as it often represents an attempt to show a correlation where there isn't one.

In any type of graph, you should be able to read all the information presented and figure out what's going on. If the presentation seems off – like if 935,575 is way lower on a graph than 327,000 – you should question if what you're looking at represents the data accurately and fairly.

CHAPTER 6: APPLIED STATISTICS IN REAL LIFE

Statistical analysis lives all around you, whether you know it or not. Nearly everything you do in your life is influenced by statistics. Remember that statistics is simply the analysis of large sets of data in order to make sense of it. Let's look at a few examples.

Imagine a mom who has a full-time job and three children. On a typical day, she makes lunches, hurries the kids onto the school bus, drives herself to work, works all day (eating lunch at her desk), makes a quick stop at the grocery store, drives home, makes dinner, helps the kids with schoolwork and baths and such, and finally puts the kids to bed. Maybe this is a dad or another caregiver instead of a mom, or maybe they split the duties. Someone, either the partner or a babysitter, was there in the afternoon to get the kids off the bus, help them with their homework, and feed them (since kids are always ravenous when they get home from school). Mom then collapses into bed, hoping to get at least seven hours of sleep before she does it again.

Hidden statistics govern every decision Mom makes during the day. She knows what time to set her alarm for because she has done this many times and knows about how much time it will take her to get herself and the kids ready. In other words, she has a sense of the average, or **mean**, amount of time she needs for each task. She packs lunches, thinking about a good balance of protein and carbohydrates in each kid's meal because nutritionists, doctors, and even governmental organizations have given her guidelines based on analyses of large sets of data.

Mom then ushers the kids out to the bus stop, knowing that the bus usually comes at 7:52. That bus schedule was made by taking several large sets of data – all of the addresses of the kids that needed to be picked up, as well as traffic patterns and street directions – and sorting them to create the most efficient bus route. Most likely, Mom isn't aware of this, but the person who coordinates the bus routes certainly is!

She takes what is usually the fastest route to work (again, based on averages) and begins her workday. The typical work hours, 9 to 5 five days a week, exist because business titan Henry Ford decided in 1926 that those hours made for the most productive factory workers (again, based on data).[xxx] Whether or not that's still true is a matter of debate, but much of our society rests on a decision made around productivity or the average amount of time people can work before their stamina and focus decline.

You get the point here. Every aspect of Mom's day, from her decision about what time to get up to how much money she spends at the grocery store, is determined by statistical analysis, either by her or others. Most people have no idea these calculations exist or that they're doing them in their heads, but we wouldn't be able to function as a society without them. (If you think the bus schedule is complicated now, imagine the bus coming haphazardly on whatever route and at whatever time the driver wants it to!)

Let's revisit Mom's workday for a minute. Henry Ford's decision to create a 9 to 5 work was based on a lot of data about productivity, including Frederick Winslow Taylor's theory of scientific management. Taylor, an engineer and manager from Philadelphia, thought that the principles of scientific investigation could be applied to factory workers and managers. His 1911 book,

The Principles of Scientific Management, laid out a theory for scientifically managing workers.[xxxi] Taylor proposed studying factory workers' minute movements and timing to find "one method and one implement which is quicker and better than any of the rest."[xxxii] He thought managers should use data – large sets of it – to analyze how to get work done most efficiently. His theory of scientific management still has an influence on how we work today. Assembly lines, where workers perform one step of a process over and over, exist largely because of Taylorism and the belief that precise, repeated actions lead to higher production.

Another way that statistical analysis influences our lives without us even realizing it has to do with something we use every single day: clothing. Averages determine clothing sizes. Once upon a time, every item of clothing was handmade for the person who would be wearing it. Eventually, once clothing manufacturing shifted to outside the home, companies needed approximate sizes to mass-produce clothing to fit many people. Those sizes were determined by taking a lot of measurements of a lot of different people and averaging them; in this way, an "average" person is fictional – it's highly unlikely that one person has exactly the average measurements in every dimension. No wonder shopping for clothes that fit well can be so frustrating!

Let's look more at this idea of averages being applied to humans. Believe it or not, averages weren't used to describe human phenomena until the mid-nineteenth century when a Belgian mathematician and astronomer named Adolphe Quetelet decided to start applying what had previously been an algorithm for astronomers to humans. Quetelet began by finding data on the chest measurements of more than 5,000 Scottish soldiers. After applying the formula for the mean, Quetelet determined that the "average" Scottish soldier had a chest size of 39.75 inches.[xxxiii]

What's fascinating about this idea of average is that it's possible not a single Scottish soldier had a chest measurement of 39.75 inches. Let's think about it in simpler terms, using the cookie example from Chapter 1. If four kids have one cookie each and the fifth kid has nine cookies, the average (or mean) number of cookies each kid possesses is three. But not a single child has three cookies. Here's another example that's even more obvious: In 2022, the average number of children per family was 1.94. Nobody actually has 1.94 children.[xxxiv]

Averages give us some information but not all the information we might need to understand a situation.

Quetelet, however, thought of the average as an ideal. He went on to compute averages of all sorts of human phenomena – marriage, divorce, and crime rates – and used them to explain societal patterns. Abraham Lincoln used Quetelet's ideas to figure out how to clothe the Union Army, finding "average" sizes for small, medium, and large soldiers.[xxxv] Next time you feel annoyed that the sweater you want doesn't fit quite right – too large in some places, too small in others – thank Quetelet and Lincoln!

The average (mean) is probably the measure of central tendency that you have heard the most about. But the other measures of center – median and mode – can also give us important information. The complete picture comes from looking at all three together.

Income inequality in the United States is a topic we hear about often on the news, in politics, and in other aspects of American life. Most of us know that this country has a very wide gap between the poor and the rich. In a 2011 article in *Vanity Fair*, economist Joseph Stiglitz claimed that the top 1% of earners in this country control 40% of the wealth.[xxxvi] This article served as one of the bases for the Occupy Wall Street movement.

What can the three measures of central tendency tell us about income inequality in the U.S.? According to census data, the mean income in the U.S. in 2021 was $97,962. That sounds pretty good, right? Taken at face value, one might think that most people in the U.S. earn about this much money, which is far from true. The median income, which is what is most often reported, was $69,717.[xxxvii] The mode isn't normally calculated but seems to be even lower than that.[xxxviii]

Remember that the median is the number that lies in the middle when you put them in order; it's the halfway mark. So approximately 50% of the people in this country earn more than $69,717, and approximately 50% earn less. The mode is the number that appears most in the data set – the number you are most likely to encounter. If you walked down the street in most cities or towns and asked people what they made, you'd hear something less than $69,717 more often than you would hear other numbers. In many ways, people think of this when they think of the "average" or typical American.

The fact that the mean income is much higher than the median and mode tells us that a concentration of wealth exists at the high end – which we know is true. The highest earners skew the data so that the mean is misleading. Think of the cookie example again. The mean of 3 cookies is misleading since one child has most of the cookies and the other children have few. Anytime the data set is skewed or has outliers – extremes at the high or low end (like the child who has 9 cookies) – all three measures of center need to be considered in order to get a more accurate read on the data.

So far, we have looked at examples of descriptive statistics in real life. Now let's look at some examples of inferential statistics. Remember that inferential statistics involves studying a large population sample to try to make inferences about the population. The study of left-handers versus right-handers described in Chapter 1 was an example of an inferential statistics study (one done poorly).

Anytime you hear a prediction about a group of people – say, what percentage of the country will vote for a certain candidate or how many people will have contracted Covid-19 by the end of the year – you are likely hearing information gathered from a sample of people and then used to predict a pattern for the entire population. Inferences about populations are made by taking data from a representative sample and applying proportional thinking to the general population. If a random, unbiased sampling shows, for example, that one person out of ten people surveyed can wiggle their ears, we could then predict that about thirty-three million people in the United States, a country of approximately 332 million people, can wiggle their ears. The same ratio (or fraction or percentage) of ear wigglers to non-wigglers – one person out of ten, or one-tenth, or ten percent – should exist in the general population as in the sample.

Political pundits use inferential statistics to predict who will win upcoming elections. Years ago, you may have gotten a telephone call asking who you planned to vote for; these days, most polling is done online, and algorithms use various data about people to make predictions. When polling, organizations need to identify a representative sample of people – a smaller subset of the population that they think reflects the beliefs of the larger population. They take the results of polling the sample and then use proportions to apply them to the larger population. This is how they can make predictions like "56 percent of the country will vote for candidate X" when they have only polled a few thousand people.

Have you ever heard a statistic about how many millions of people watch a certain television show and wondered how anyone knows what you're watching? Nielsen, the research group that publishes these statistics, carefully selects a panel of several thousand households to conduct their studies. From the sample's viewing habits, they estimate the viewing habits of the nearly 121 million homes with televisions in the U.S.[xxxix]

This process doesn't always work, though. As in the study of left-handed people, sometimes hidden factors are at play that can influence the outcome of the study. Maybe hidden biases existed, the methodology was flawed, or the sample didn't consider some factors. This happened on a large scale with the presidential election of 2016.

In the days leading up to Election Day, news outlets showed the election going decisively to Hillary Clinton, the Democratic candidate. Some sources even went so far as to say there was a 99% chance that she would win.[xl] Democrats felt confident, knowing that these predictions were based on samples and data. But Election Day came, and while the popular vote went to Clinton, the electoral college went to Donald Trump, the Republican candidate. Pollsters were shocked; how had they missed this? One statistician and professor from Princeton, Sam Wang, ended up eating a live cricket on CNN as a consequence of his over-confidence.[xli]

Statisticians have tried to explain how they got the 2016 election so wrong. There is no clear answer, but rather several possible factors. Some suspect that a larger percentage of people who voted for Trump didn't respond to polls, leading to a hidden bias in the samples. Another possibility is that different people showed up to vote than those who had been polled – another error in the sample.[xlii] What this example illustrates is, again, that statistics applied to large populations tell us *something*, but not everything. It also shows us how difficult it is to make accurate predictions about an entire population based on inferences from a sample.

Inferential statistics is about *inferences*, not absolutes. We gather data from smaller samplings because polling the entire population – the country, the world, every single voter, etc. – would be too difficult. Listen for phrases like "In a small study of ___ people…" or "Based on polls, we predict that…." The questions to ask are how the sampling was done and whether the study could be replicated in another place, at another time, or on a larger scale to get the same results. Hidden biases are sneaky; they can even fool top scientists, as we saw with the left-handed study.

CHAPTER 7: VISUAL DISPLAYS: TELLING STORIES THROUGH IMAGES

Take a look at the following graph:[xliii]

BUS RIDERSHIP IN MIAMI, MINNEAPOLIS, ATLANTA AND PORTLAND, ORE.

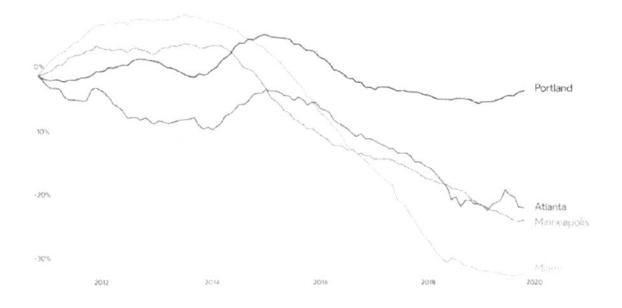

What do you think the graph shows? What is it trying to tell us? Every graph tells a story, and really good ones make the story both accessible and interesting. According to Mark Nelson, President and CEO of Tableau, "Data storytelling is the ability to distill data down to the insights you want and present those insights as a story."[xliv] In this chapter, you will learn how to read the story a data display is telling.

In the graph above, we see bus ridership in four major cities between about 2010 (note that the beginning of the x-axis isn't labeled) and 2020. The lines on the graph, each representing a city, increase and decrease a bit, but all four of them decline in that time. All of them, except for Portland, began a sharp decline around 2014 or 2015. So, what's the story of this graph? Bus ridership in these four cities has largely declined since 2010.

Data can be displayed in a number of different ways, from graphs to charts to infographics. Each of these ways emphasizes different aspects of the data and affects a viewer differently. The type of data and the conclusions drawn help scientists determine what kind of display to use. Indeed, the wrong type of display, whether accidental or purposeful, can cause viewers to draw conclusions not borne out by the data. In this chapter, we'll break down the most popular kinds of visual displays and what questions to ask yourself when viewing them.

To begin making sense of a visual display of data, you need to look at all of the information provided. Think of the display as a puzzle that you are trying to figure out, and ask yourself the following questions:

- What is the title of the graph?
- If there are axes, what does each axis represent, and what is the scale of each axis?
- Can I pick one data point and figure out what it represents?
- What is the shape of the data? Does it show a certain correlation or trend?
- Is there any additional information on the graph or display? If so, what does it tell me?
- What story is this graph or display telling?

Line Graphs

Line graphs are the most straightforward type of graph on the coordinate plane. They often show change over time or another relationship between two variables, one of which is dependent on the other. Line plots have an x-axis (horizontal) and a y-axis (vertical) and should also have a title and label for each axis. Pay attention also to the units that each axis is measured in; this tells you what two measurements the graph is showing the relationship between. Each axis's scale should also be clear and proportional, without any obvious gaps or leaps.

Take a look at the following graph. [xlv] Start by reading the title and axis labels to get a sense of the data it shows.

INCOME GAINS WIDELY SHARED IN EARLY POSTWAR DECADES BUT NOT SINCE THEN

Real family income between 1947 and 2018, as a percentage of 1973 level

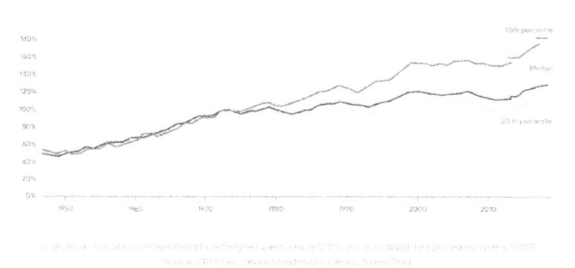

Notice the scale and units of the axes: the y-axis starts at zero and is marked in increments of twenty percentage points. The x-axis shows time, starting in 1947 and ending in 2018, in increments of one year (with each decade labeled). Now look at the three different gray lines. Using the line labels along with the graph's title, we can deduce that the "real family income" of the 95th percentile of earners grew faster than that of the 20th percentile of earners. If we look at the shape of each line, we can see that the median income mostly paralleled the 20th percentile income, which tells us that they grew and fell at about

83

the same rate. The 95th percentile income, however, grew much more steeply between about 1980 and 1990 and then again in the late 90s. So, what's the story this graph is trying to tell us? The broad outline of what it tells us is that the income of higher earners grew much more than that of lower earners between about 1980 and 2018. The story can often be summarized by the title of the graph (as it is in this case) or the headline of the article that the graph appears in.

Scatterplots

Scatterplots are used when you have a set of discrete data points, and you're trying to show an association between them. Each point on the graph represents bivariate data or data that has two inputs. For example, the scatterplot below shows the association between foods that average Americans think are "healthy" and foods nutritionists deem healthy.[xlvi] One input, measured on the x-axis, is foods Americans deem healthy; the other, measured on the y-axis, is foods nutritionists deem healthy.

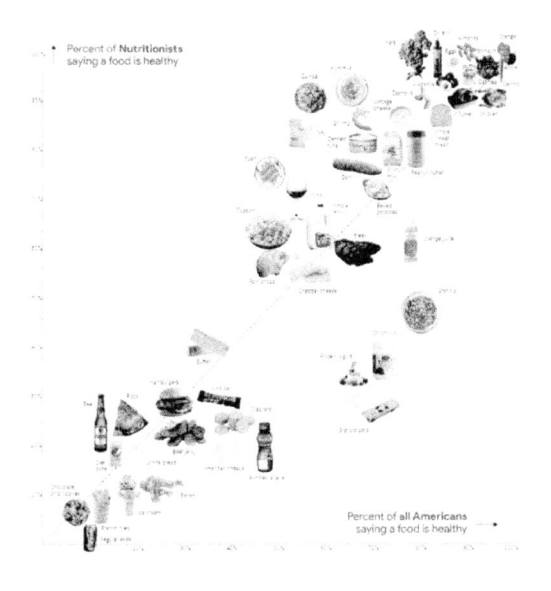

Each point on the graph – which, in this case, is represented by a picture of the food – represents one data point or measurement. Look at cheddar cheese, for example. Based on the coordinates of cheddar cheese, we can tell that about 55% of average Americans (the x-axis), and 55% of nutritionists (the y-axis) think it is healthy. Notice that there is a diagonal line going right through cheddar cheese. That line is the graph of y=x, which has a slope of 1, meaning there is a 1:1 correlation. In other words, foods that fall on that line are deemed healthy by an equal percentage of Americans and nutritionists. It looks like we all agree that turkey and spinach are quite healthy and that chocolate cookies are not. Foods that fall below the diagonal line are deemed healthy by more Americans than nutritionists, and foods located above it are deemed healthy by more nutritionists.

Scatterplots can show a strong positive correlation, a strong negative correlation, no correlation, and everything in between. The association can be linear, as in the above graph, exponential, quadratic, inverse, or another shape. If there is a strong correlation, an equation can model the data, and making predictions about data we have not yet collected is easier. Scatterplots also show us clusters and outliers in the data.

Look again at the food graph above. We can see two clusters of images, one towards the bottom left and one towards the top right. These are foods that most of us agree are either unhealthy or healthy. But notice those few items that are not in the clusters and not near the diagonal line – granola, granola bars, orange juice, and a few other things. These are the outliers that stand apart from the rest of the data. Based on where they fall on the graph (higher on the x-axis, lower on the y-axis), we can tell that a higher percentage of average

Americans than nutritionists think these foods are healthy.

Bar Graphs and Histograms

Bar graphs show categorical data in easy-to-see bars lined up next to each other (usually, but not always, along the x-axis). Bar graphs also have a title and two axes, which may show various data categories within each bar. Take a look at the following graph, which shows budgets in various countries by type of technology[xlvii]:

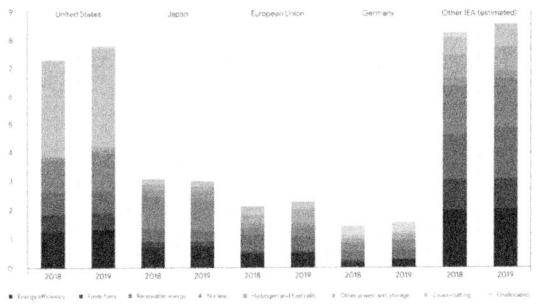

2018 AND 2019 BUDGETS BY TECHNOLOGY IN SELECTED IEA COUNTRIES AND THE EUROPEAN UNION

This graph has a lot going on, which means we need to study it for a few moments before we can draw conclusions. This is a "stacked" bar graph, meaning that each bar has more than one category within it (the categories are stacked). Make sure you notice the key at the bottom that tells you what the stacks in each bar represent. Also, notice that the x-axis is broken into 5 sections showing one country's budget in 2018 and 2019. The graph asks us to compare across countries and also across years within each country.

Once you've understood how the graph is laid out and what the labels mean, ask yourself what you notice about what is shown. You may notice that the U.S. has a much higher percentage of cross-cutting (second gray from top down) than the other countries do; this means that the U.S. devotes a higher percentage of its budget to "cross-cutting" technology than the other countries shown here. You may also notice that Germany's bars in both years are lower than any other country, meaning they devoted much less money towards any of the technologies mentioned.

Histograms are similar to bar graphs but group the data into ranges or "bins" to try to make more sense of it. Look at the following graph of used car sales by mileage in 2021[xlviii]:

TOTAL OF USED CARS BY MILEAGE

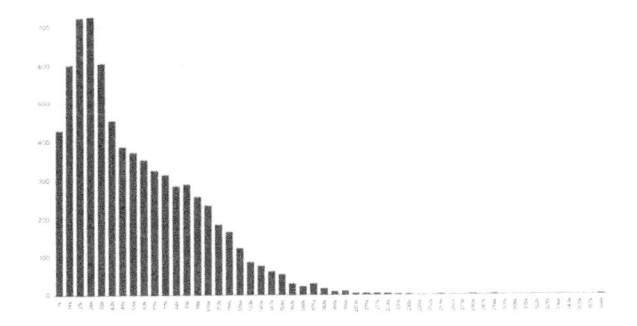

We know that most used cars wouldn't have exactly the number of miles shown above at sale time. The labels along the x-axis – the bins – show groupings of data to help us make sense of it. We can see that the largest percentage of cars sold had between about 14,000 and 35,000 miles on them. By "binning" the mileage, we can make better sense of the data and see the distribution of the data more clearly.

Histograms are great for looking at the shape of the data. Notice that the data in the above graph is clumped towards the left side, trailing off on the right as the cars have greater mileage. We call this data skewed right, meaning (somewhat counterintuitively) that the data points lie more on the lower or left-hand side of the graph. This makes sense when you think about used cars – most people don't want to buy cars with a lot of mileage already. Compare this to data on the year of the used cars that were sold, which is skewed left:

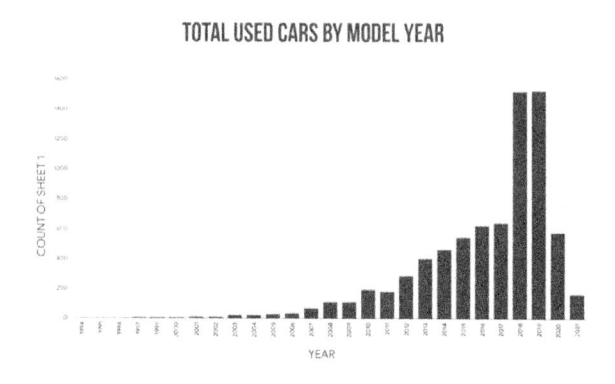

TOTAL USED CARS BY MODEL YEAR

Again, this makes sense based on common knowledge: most people would rather buy used cars that are only a few years old.

<u>Circle Graphs</u>

Circle graphs, also known as pie charts, are great for comparing different categories. The full circle represents 100% of the population being surveyed; each slice within the circle represents the percentage of 100 that that category of responses comprises. The largest sliver represents the category with the highest percentage; the smallest sliver represents that with the lowest percentage. Let's look, for example, at what kinds of cheese Americans ate in 2012[xlix]:

SHARE OF U.S. CHEESE AVAILABILITY PER CAPITA, 2012

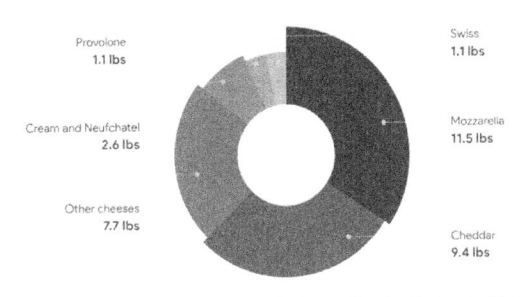

Provolone
1.1 lbs

Swiss
1.1 lbs

Cream and Neufchatel
2.6 lbs

Mozzarella
11.5 lbs

Other cheeses
7.7 lbs

Cheddar
9.4 lbs

We can see that dark blue makes up the largest chunk of the circle, meaning Americans consumed more mozzarella in 2012 than any other type of cheese. This makes sense when you think about how much pizza Americans consume!

Circle graphs are most appropriate when you are comparing shares that add up to a whole or 100% of whatever was surveyed. Let's say you are a teacher with 20 students in your class. You survey them about what their favorite sport is and get these results:

5 students: basketball

4 students: baseball
8 students: football
3 students: undecided

We can rewrite these as a percentage of the class:

25% (5/20): Basketball
20% (4/20): Baseball
40% (8/20): Football
15% (3/20): Undecided

These results should add up to 100% because you got an answer from every student in your class. If you are making a circle graph and aren't sure exactly how to mark 20% of the whole circle, think back to high-school geometry. A circle has 360 degrees, so you can set up a proportion to figure out how many degrees each percentage (or each fraction) should comprise.

$$\frac{20}{100} = \frac{?}{360°}$$

By solving the proportion, we find that 20% (20/100) of 360° is 72°, so we would make the "Baseball" section take up 72° of the circles.

(Most digital tools will do this step for you!) Now we can draw the circle graph:

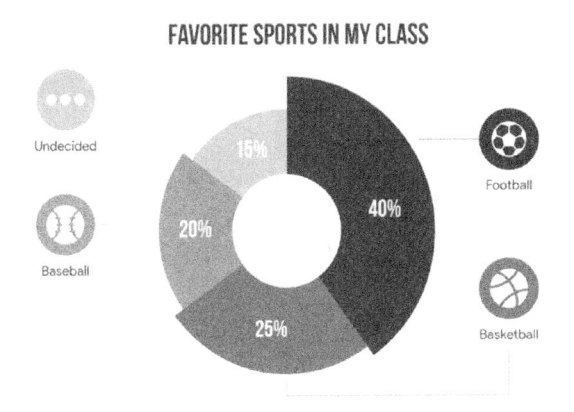

FAVORITE SPORTS IN MY CLASS

- Undecided: 15%
- Baseball: 20%
- Basketball: 25%
- Football: 40%

We'll take a look in Chapter 9 at how circle graphs can be misused in an attempt at data manipulation.

Other Visual Displays

Line graphs, scatterplots, bar graphs, histograms, and circle graphs may be the most common visual displays you'll encounter, but many other types of displays can send powerful messages. Heat maps, for example, show the relative frequency of the thing being measured. Take, for example, this heat map from the

Washington Post showing "literally every goat in the United States":

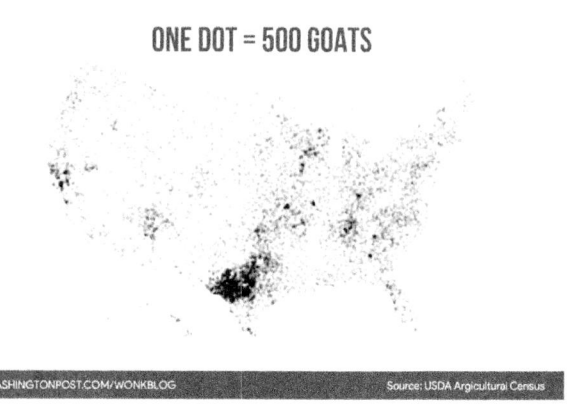

ONE DOT = 500 GOATS

WASHINGTONPOST.COM/WONKBLOG Source: USDA Argicultural Census

    The darker areas are where more dots are clustered, meaning more goats live there. Texas has the rest of us beat in terms of goat population!

    Even more exciting than heat maps and other graphs are the types of visual displays that many online news outlets have created in the past decade. Many of these graphics exist online and are interactive, allowing the user to explore them in great depth. Some of them are optimized for phones, where you can scroll through and get more and more information as you scroll. Technology has opened up a whole new world of data displays, and there is no shortage of data to explore.

CHAPTER 8: MISINTERPRETATION OF STATISTICS: 5 COMMON PITFALLS

As you learned in the previous chapters, it is easy to get statistics wrong. Getting statistics *right* is really hard, which is why there are trained experts whose job is to sort through data! And even those trained experts can sometimes make mistakes. This chapter will look at five common pitfalls in statistics and how you, the average consumer, can recognize them.

Pitfall #1: Ignoring scale

A basic understanding of percentages and proportions is helpful when interpreting data. Sometimes what seems like a big change actually means very little when you look at the overall data, and sometimes a tiny change is quite significant. Let's look at an example of cats and dogs to understand this point further.

According to most sources, domestic cats usually weigh between about five pounds and fifteen pounds (with a few exceptions of very large breeds). Domestic dogs usually weigh between about five and one hundred fifty pounds, again with a few extremes omitted. If we take these numbers as true, the range of cat weights is about ten pounds (the difference between fifteen and five), and the range of dog weights is about one hundred forty-five pounds. There is much more variability in dog sizes than in cat sizes!

Let's say you own a ten-pound cat and a one-hundred-thirty-pound Bernese Mountain Dog. You take them to the vet for check-ups, and the vet tells you that they've each gained two pounds since they were weighed last year. Then the vet tells you you need to put your cat on a diet. "But why?" you might exclaim. "He only gained two pounds! Why would he have to go on a diet, but the dog, who eats everything in sight, including our sofa, wouldn't?" The answer here has to do with relative size. Your dog gained only a little over 1% of its body weight by gaining two pounds. Your cat, however, gained 20% of its body weight at that same time. In relation to their sizes, the cat gained a much higher proportion of weight than the dog did.

Another example of small but meaningful changes is how the Securities and Exchange Commission (SEC) investigates white-collar crimes. The SEC has tools that constantly monitor the sales patterns of company shares. Let's say they notice a number of sales from people connected to a company before a major event, say, a collapse or announcement of a merger. The SEC uses high-tech data analysis to determine if these sales are statistically significant or follow the expected pattern. If the sales are deemed statistically significant, the SEC might launch an investigation into insider trading.[1]

On the other extreme, some people think that everything is statistically significant when small changes often aren't. This happens most easily with studies that include a small sample size. When looking for statistical significance, data analysts calculate the p-value, which we looked at in greater depth in Chapter 4. Remember that the lower the p-value, the more statistically significant something is deemed to be.

Let's look at the case of Ivermectin and its promise to treat Covid-19. In the spring of 2020, as the new coronavirus tore through populations and threatened to kill millions globally, a small laboratory study from

Australia reported that Ivermectin, a deworming medication, killed the Covid-19 virus. Some, including the U.S. president, began to hail Ivermectin as a miracle drug that could end the pandemic. However, a large-scale study involving approximately 1600 people concluded that Ivermectin did not work on Covid-19 any better than a placebo.[li] According to the University of Kansas Medical Center's report on the study, "Researchers found that the median recovery time for those taking Ivermectin was 12 days, and those on the placebo was 13 days. There were 10 hospitalizations or deaths in the ivermectin group and nine in the placebo group. But these differences failed to be statistically significant, leading researchers to their conclusion that "these findings do not support the use of ivermectin in patients with mild to moderate COVID-19."[lii] There *was* a small difference in the reported outcomes of people who took Ivermectin versus those who took a placebo, but the difference wasn't statistically significant. People who didn't understand statistical significance might have concluded that the study proved Ivermectin worked.

Pitfall #2: Looking at the wrong measure of center

When you're interpreting statistics, it's important to know the distribution of the data, how widely varied the data points are, and where the bulk lie. A "normal" distribution is shaped like a bell curve, with most points falling in the middle of the range and then high and low points tapering off on each end. In the example of the average amount earned by U.S. families cited in Chapter 1, we can see that outliers – data points at the extremes, like the top 1% of U.S. earners – can skew the results, making the mean less, well, meaningful. The data distribution tells us which measure of center is most useful.

Imagine you are taking a course in college. For the final exam, you get an 89%. Not bad, you think to yourself. You learn that the class average was 92, and you feel satisfied that you got pretty close to that average. But you soon learn that every student in the class except for two aced the test and got 100% on it. The other two scores were you, at 89%, and a student who missed half of the semester and earned a 65%. Suddenly you don't feel as good about that 89. In this case, your score and the 65 were outliers, with the rest of the data clustered at 100. Because of these two outliers, the mean gives you a false understanding of the situation. More helpful measures of center would be the median or the mode, which would both be 100 in this case. If you had heard that 100 was the median, you would have understood that more than half the class scored 100, and you might have been able to put your 89 more in context.

On the flip side, imagine that you learned that most of the class earned scores in the high 70s or low 80s, with just a handful scoring 100. Because of those outlying high scores, the class average seemed artificially high. Once again, the median might have given you a better sense of how you did in comparison to others, and you might have felt even better about your grade.

Pitfall #3: Confusing Correlation with Causation

Mixing up correlation with causation is a big pitfall, and it's also one that leads to some of the most extreme conclusions. Correlation simply means that two (or more) phenomena are linked in some way; maybe they occur at the same time, or they fluctuate together. Causation means that one actually causes the other. Correlations exist all over the place and usually have nothing to do with causation. In fact, there are some pretty eye-opening correlations out there that people have discovered.

For one, did you know that there's a correlation between the number of films Nicholas Cage has appeared in each year and the number of people who drowned by falling in a pool? A man named Tyler Vigen has found a number of interesting correlations like this.

NUMBER OF PEOPLE WHO DROWNED BY FALLING INTO A POOL
correlates with
FILMS NICOLAS CAGE APPEARED IN

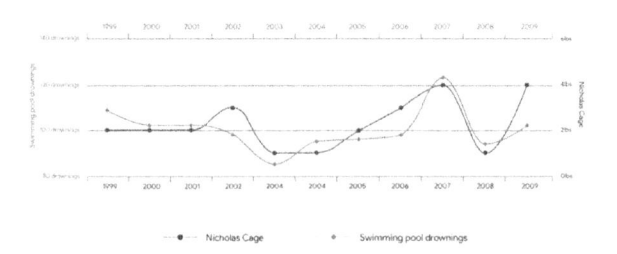

Nobody would ever think one of these phenomena causes the other because that's clearly a ridiculous idea. Vigen has a whole website devoted to "spurious correlations." Apparently, there's a nearly perfect correlation between computer science doctorates awarded in the U.S. and the total revenue generated by arcades![liii]

The correlations Vigen has found are obviously ridiculous. But many correlations are more plausible than these, and often scientists even assume that correlation is causation. This is what happened with studies of childhood vaccines and autism. The doctor who originally publicized the link between vaccines and autism had simply noted that – a link. (He also was later found to have falsified some of his data and had his medical license revoked.) Yet fearful parents heard about this link and assumed it signaled causation. Despite numerous large studies showing absolutely no causation between the two, many still believe the myth that vaccines cause autism.[liv]

Another example with more serious implications exists in the public health sphere. In recent years, public health organizations have decried "the obesity epidemic" and spent millions of dollars attempting to fight obesity. But there is still no causal link between obesity and increased health risks. There is a *correlation* between obesity and health risks (such as diabetes and heart disease), but scientists have not yet proven that obesity *causes* poor health outcomes. According to the *Lancet*, "Causal links between obesity and the risk of disease are not so simple. Obesity results from a mix of factors such as diet or physical activity, embedded in a causal web of environmental and socioeconomic determinants, which directly and specifically affect the risk of obesity-related diseases."[lv]

Most superstitions exist because people confuse correlation with causation. Maybe your favorite baseball team wins two games in a row while you're wearing red socks. If you're prone to thinking superstitiously, you might notice this connection and decide that the red socks caused the team to win. Now you'll be making sure those red socks are clean before every baseball game for the rest of the season.

Pitfall #4: Failing to see biases

We discussed a couple of examples of hidden biases in the first chapter of this book. Hidden biases are sneaky because you may think that a study was conducted perfectly and that the data tell an obvious story. But often, there are complications to the story that we can't see; it takes an outside look or a thoughtful questioner to realize when these hidden biases are present.

Let's look at a few real-life examples. About a decade ago, the city of Boston released an app called Street Bump in an effort to detect where the potholes were around the city so that they could be fixed. The app allowed users' phones to send data when they encountered potholes while driving. The problem, at least when the app was built, was that many residents, particularly in poorer areas of the city, didn't own or use smartphones while driving. The information being gathered was biased towards the areas of the city where residents used their phones as they drove.[lvi]

There was hidden bias in the studies done that predicted Hilary Clinton as the winner of the 2016 election by a large margin. For various reasons, the sample did not accurately reflect the people who came out to vote on Election Day that year. Biases are especially sneaky when it comes to political predictions.

Another famous example comes from the 1936 presidential race, when the *Literary Digest* polled its ten million subscribers and predicted that Alfred Landen (Republican) would win by a landslide over Franklin D. Roosevelt (incumbent Democrat). Turns out the magazine was wrong, and many have surmised it's because their subscribers were overwhelmingly Republican.[lvii] There are so many potential biases, in fact, that there is a catalog of biases describing all the different ways studies can be biased.[lviii]

Pitfall #5: Getting causation backwards

Data scientists are well attuned to the problem of reverse causality, so you won't see this in the news that much, but it still could affect your beliefs. Reverse causality can be seen as a subcategory of confusing correlation and causation; you know that two phenomena are correlated, but you incorrectly label one as the cause of the other.

In reality, there are many correlations in which the cause and effect are unclear. Take, for example, statistics on smoking and depression. There is a strong correlation between smoking and depression. You might be tempted to think – and some media have actually reported – that smoking leads to depression. Numerous studies, however, have not proven the causal direction.[lix] We don't know yet if smoking causes depression or if people who are already depressed tend to smoke.

It is so easy to accidentally reverse causality, in fact, that many public health experts now rely on a set of nine principles called Hill's Criteria for Causation, named after statistician Sir Austin Bradford Hill. In 1965, Hill proposed these criteria for evaluating epidemiological associations. While they are not universally used, it may be helpful to ask yourself these questions when you encounter an association between phenomena, and you're not sure what the causality is:

1) Strength of association (how strong is the association between the data?)
2) Consistency (will the same results be found if different people replicate the study at different times?)
3) Specificity (how specific is the association?)
4) Temporality (did the effect occur after the cause?)
5) Biological gradient (is there a correlation between the amount of exposure and the amount of disease?)
6) Plausibility (Is there a plausible way that the cause would lead to the effect?)
7) Coherence (Is there consistency between laboratory and epidemiological findings?)
8) Experiment (Does intervening or experimenting on the association lead to changes in the findings?)
9) Analogy (are there similarities between the observed associations and other associations?)[lx]

Some of these questions apply to certain situations better than others. For example, if we ask ourselves if Nicholas Cage appearing in movies really causes drownings, we'll see a strong association but very little plausibility.

CHAPTER 9: DATA MANIPULATION AND THE POWER OF GRAPHS

Did you know that over 80% of dentists recommend Colgate toothpaste? You may have heard that in advertisements in 2007 and picked Colgate over Crest the next time you went to the drugstore. It turns out four of five dentists did recommend Colgate, but not in the way the ads led you to believe, and the Colgate-Palmolive company was censured for this ad campaign.[lxi]

The company's survey asked dentists which toothpaste brands they recommended, and Colgate was just one of the options presented. Participants in the survey named a number of brands, and it turns out that some of the other brands were recommended nearly as much as Colgate was. So yes, 80% of dentists did recommend Colgate, but that same 80% might have also recommended Crest or Sensodyne. The ad campaign was deemed misleading since most viewers would interpret that claim to mean 80% of dentists recommend Colgate *over* other brands.[lxii]

As we've seen in a few examples, even if a study has been done well and the numbers presented are accurate, data can be wildly manipulated to try to convince the audience (usually you) of something. One common trick is to zoom in on one of the axes to make an effect seem larger than it really is. *The Times*, a UK newspaper company, did this in a 2006 article about how much more popular they were than other news sources. They published this graphic along with the article:

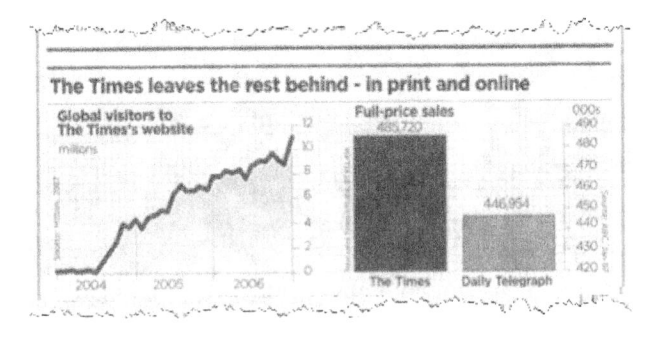

The Times leaves the rest behind - in print and online

A quick look at the graph on the right gives us the impression that *Times* sales were nearly double *Daily Telegraph* sales in whatever time this represents. If you look more closely at the numbers along the y-axis, you'll see that they've zoomed in to show just 420,000 to 490,000. They've cut off all the numbers below 420,000 because they want you to think that a relatively small difference in sales – about 40,000 people, or less than ten percent of either company's base – is enormous.[lxiii] As with the Planned Parenthood graph, the numbers are true, but the way the graph has been drawn makes you believe a story that the numbers alone don't tell.

This effect, where one axis shows only a portion of the total numbers, is called omitting the baseline. The baseline number is almost always zero; the y-axis should reflect that. In the case of the *Times*, had the y-axis included all of the numbers from 0 to 420,000, we would have seen that the difference in sales between the two papers is not that large. But, of course, that's not what the *Times* wanted us to see.

Another way to mislead people is to leave key information out of the graph. Take the example of Nielsen – the company we normally trust with television ratings – publishing data that made it look like the Wii was headed toward oblivion. Gaming sites latched onto this data and published the following graph:

6-MONTH TREND - ACTIVE USER PERCENTAGE

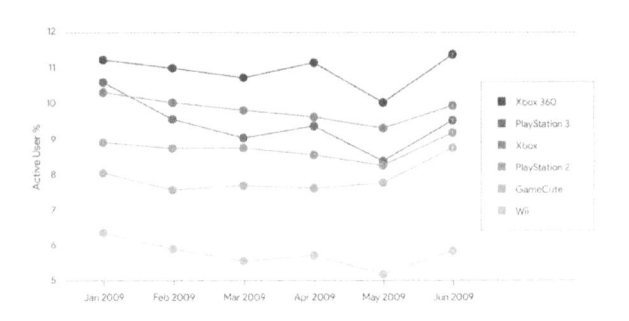

Wow, the Wii is pretty unpopular compared to other platforms, right? That's a logical conclusion based on this graph. What they didn't tell you is the number of users of each, that this graph shows the percentage of active users *out of the total number of users for each platform*, and that there were way more Wii users than users of any other platform. Blogger Lee Evans explains it like this:

"Take a look at the top number on the graph. Only 11% of 360 owners actively use them.
360 and only 10% of PS3 owners actively use their PS3. Now, let's do a little math.

There are 50 million Wii owners. 6% of that number is 3 million.

There are 30 million 360 owners. 11% of that number is 3.3 million.

There are 20 million PS3 owners. 10% of that number is 2 million."[lxiv]

So, there were fewer PS3 players than players of the Xbox 360 or the Wii.

Another way that graphs can purposely be misleading is if the creator of the graph uses the wrong type of graph for the information. The type of graph we have looked at so far, where there are two axes and results are compared in the coordinate plane, is most often used to compare two or more things. We can easily see where each item measures up against the other and against the scale marked on the y-axis. News outlets have become adept at creating other types of displays that may or may not be appropriate for the data.

Let's look, for example, at circle graphs. As we discussed in Chapter 7, circle graphs, by definition, should represent 100% of responses, and they can be misleading if the responses displayed don't add up to 100%. This is most likely to occur if survey respondents are allowed to pick more than one preference. This graph appeared on Fox News during the 2012 election.[lxv]:

2012 PRESIDENTIAL RUN
GOP CANDIDATES

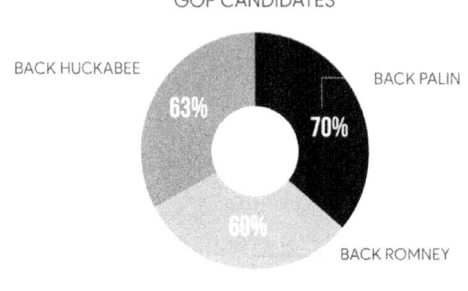

BACK HUCKABEE 63%
BACK PALIN 70%
BACK ROMNEY 60%

SOURCE: OPINIONS DYNAMIC

The problem with this graphic isn't that the data wasn't true. The problem is that people were allowed to pick more than one candidate they backed, so the results don't add up to 100%. The " pie " slices in the pie chart don't mean anything since they do not show a proportional relationship between responses. This graph would make more sense if it were a bar graph that made it clear that the people surveyed could pick more than one response.

A more sinister phenomenon is when the actual data is manipulated. One of the most influential examples of this comes from a man named Brian Wansink, once a Cornell professor, well-known social scientist, and author. Wansink ran the Food and Brand Lab at Cornell, which focused on people's choices around food. He was particularly concerned with obesity and weight loss. While he directed the USDA's Center for Nutrition and Promotion from 2007 to 2009, the USDA created numerous influential recommendations for Americans' diets, including the 2010 Dietary Guidelines for Americans. Much of what you have heard about Americans' eating habits, including that "bottomless bowls" will make you eat more and that the portion sizes in The *Joy of Cooking* have increased over time, came from Wansink.[lxvi]

In a 2016 blog post, Wansink described the work he and a graduate assistant did to analyze data that led to several published research papers. The data mining he described set off alarm bells for other scientists, who questioned his methods of analyzing and re-analyzing data until he found something useful. Most scientists start with a hypothesis, then collect data to see if their hypothesis bears out. He collected a ton of data, then scoured it until he could make a point – a point that he knew would get him media attention, given his prominent position.[lxvii]

Having noticed something fishy happening with one data set, data analysts and scientists began looking at Wansink's other studies. Since 2017, Wansink has had eighteen papers retracted, fifteen corrected, and seven received an "expression of concern." Cornell decided what he had done amounted to scientific misconduct and was forced to resign.[lxviii]

While Wansink is an extreme example, he is surely not the only scientist to succeed (at least for a while) in publicizing manipulated data results. One major pitfall of studies on nutrition is that many, if not most, of them rely on self-reported data. Sometimes the self-reporting comes from a food diary and sometimes from a simple recall of what a person thinks they ate.[lxix] If you were asked how many times you ate blueberries in the past six months, could you answer accurately? What about if you were also asked what the serving size was? This kind of self-reported data is notoriously unreliable, and yet much of what we hear about nutrition comes from studies that use it.

CHAPTER 10: CONCLUSION

Now that you've reached the conclusion of this book, you might be thinking that you can't trust anything you see. Studies can be invalid and unreliable! Graphs can be misleading! Data can even be manipulated or made up! While all of these statements are true, they should not be your takeaway.

This book aims to teach you that statistics are both difficult and incredibly important to get right. Scientists, data analysts, and researchers receive thorough training in how to conduct scientific studies. They spend months crafting an appropriate question and sometimes years figuring out how to collect accurate data. They encounter errors, discover mistakes, and have to discard their data and start again. This is one of the reasons it was so remarkable that biotech companies around the world crafted, tested, and mass-produced a vaccine for Covid-19 so quickly.

Now that you know how hard it is to create and report accurate statistics, you should have a more discerning eye when you encounter statistics in your life. You will be able to recognize when a graph is misleading; you'll know to take the results of a study with a grain of salt if they seem improbable; you'll have the understanding to discern valid results that can be replicated from invalid and unreliable ones.

You also may have gained an appreciation for the art of statistical display. Choosing the right way to represent data is no easy feat, which is why many online publications now have entire teams dedicated to graphical displays. *The New York Times* even has a suggested lesson for teachers to use each week in a column called "What's Going on in This Graph?"[lxx]

Data analysts' and graphic artists' work is critical in the age of information and misinformation. A simple decision about what color to use in a data display can lead people to draw vastly different conclusions. Take, for example, this map published by the Pew Charitable Trust in 2018:

POPULATION GROWTH

Population growth slowed last year in some of the nation's most expensive countries, like choose in California's Silicon Valley, and picked up in more affordable countries in the Sun Belt. Hover over countries for details.

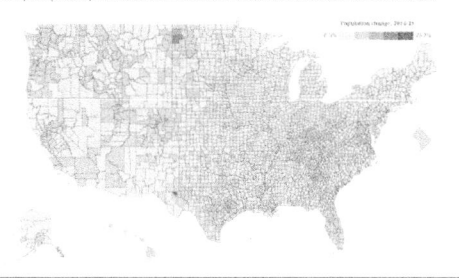

Stateline data visualization, April 2016 | Source: U.S. Census Bureau

As Washington Post columnist Christopher Ingraham describes in an article titled "The Dirty Little Secret that Data Journalists Aren't Telling You," the folks at Pew were attempting to show population growth and decline across the US in 2014-2015, according to census data. However, by choosing to show growth and decline in shades of brown and categorizing the numbers the way they did, they ended up with a boring visual display that's hard to make much sense of. Where did the population actually decline in the US, and where did it grow? We can't easily tell from this map.

Ingraham took another stab at a map using the same data. He categorized the numbers differently, so there was a clear visual difference between growth and decline, and he made the colors more distinct. This is what he came up with:

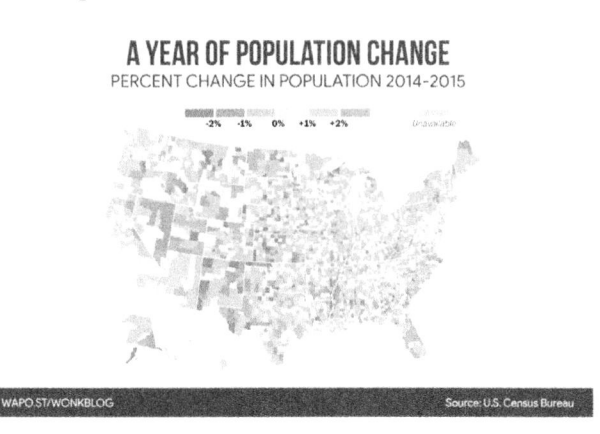

A YEAR OF POPULATION CHANGE
PERCENT CHANGE IN POPULATION 2014-2015

Ingraham's map is much easier to make sense of. It grabs our attention and tells a clear story about where and by how much the US population changed in 2014-15.

Ingraham's purpose wasn't to denigrate the Pew Charitable Trust but rather to show how much thought goes into visual displays of data. "Visualizing data is as much an art as a science," he wrote, "...Numbers carry a veneer of authority and objectivity that words can seem to lack. But communicating with numbers is, in many ways, just like communicating with words. You make decisions about what to emphasize and what to downplay and about how to convey a full understanding of the subject at hand."[lxxi]

If you've gotten this far in the book and are still interested, there is a whole world of statistics out there waiting to be analyzed. Just remember to plan each step of the way so that the data you present to the world is accurate and reliable and tells a true story about the phenomenon you studied.

Thank you for reading this book.

Respectfully,
    A. R.

Before You Go…

I would be so very grateful if you would take a few seconds and rate or review this book on Amazon! Reviews – testimonials of your experience - are critical to an author's livelihood. While reviews are surprisingly hard to come by, they provide the life blood for me being able to stay in business and dedicate myself to the thing I love the most, writing.

If this book helped, touched, or spoke to you in any way, please leave me a review and give me your honest feedback.

CLICK HERE TO REVIEW

Also, don't forget to pick up your gift, The Art of Asking Powerful Questions. Visit www.albertrutherford.com for further details.

Thank you so much for reading this book!

About the Author

Albert Rutherford

We often have blind spots for the reasons that cause problems in our lives. We try to fix our issues based on assumptions, false analysis, and mistaken deductions. These create misunderstanding, anxiety, and frustration in our personal and work relationships.

Resist jumping to conclusions prematurely. Evaluate information correctly and consistently to make better decisions. Systems and critical thinking skills help you become proficient in collecting and assessing data, as well as creating impactful solutions in any context.

Albert Rutherford dedicated his entire life to find the best, evidence-based practices for optimal decision-making. His personal mantra is, "ask better questions to find more accurate answers and draw more profound insights."

In his free time, Rutherford likes to keep himself busy with one of his long-cherished dreams - becoming an author. In his free time, he loves spending time with his family, reading the newest science reports, fishing, and pretending he knows a thing or two about wine. He firmly believes in

Benjamin Franklin's words, "An investment in knowledge always pays the best interest."

Read more books from Albert Rutherford:
Advanced Thinking Skills
The Systems Thinker Series
Game Theory Series
Critical Thinking Skills

Resources

2018 and 2019 budgets by technology in selected IEA countries and the European Union – Charts – Data & Statistics - IEA. (n.d.). IEA. https://www.iea.org/data-and-statistics/charts/2018-and-2019-budgets-by-technology-in-selected-iea-countries-and-the-european-union-2

Aschwanden, C. (2021, March 8). *You Can't Trust What You Read About Nutrition*. FiveThirtyEight. https://fivethirtyeight.com/features/you-cant-trust-what-you-read-about-nutrition/

Average children per family U.S. 2022 | Statista. (2022, December 13). Statista. https://www.statista.com/statistics/718084/average-number-of-own-children-per-family/#:~:text=The%20typical%20American%20picture%20of,18%20per%20family%20in%201960.&text=If%20there's%20one%20thing%20the,is%20known%20for%2C%20it's%20diversity.

Barnes, B. H. (2013, September 7). *Do left-handed people really die young?* BBC News. https://www.bbc.com/news/magazine-23988352

Baykoucheva, Svetla (2015). *Managing Scientific Information and Research Data.* Waltham, MA: Chandos Publishing. p. 80. ISBN 9780081001950.

Bayes for days: What to do with signal | Mawer Investment Management Ltd. (n.d.). https://www.mawer.com/the-art-of-boring/blog/bayes-for-days-what-to-do-with-signal

Bayes Theorem Application in Everyday Life : Networks Course blog for INFO 2040/CS 2850/Econ 2040/SOC 2090. (2018, November 19). https://blogs.cornell.edu/info2040/2018/11/19/bayes-theorem-application-in-everyday-life/

Beers, B. (2023, March 28). *P-Value: What It Is, How to Calculate It, and Why It Matters.* Investopedia. https://www.investopedia.com/terms/p/p-value.asp

Biases Archive. (n.d.). Catalog of Bias. https://catalogofbias.org/biases/

Calzon, B. (2023, March 1). *Misleading Statistics – Real World Examples For Misuse of Data*. BI Blog | Data Visualization & Analytics Blog | Datapine. https://www.datapine.com/blog/misleading-statistics-and-data/

Caporal, J. (2023, February 1). *Are You Well-Paid? Compare Your Salary to the Average U.S. Income*. The Motley Fool. https://www.fool.com/the-ascent/research/average-us-income/

Chiolero, A. (2018). Why causality, and not prediction, should guide obesity prevention policy. *The Lancet. Public Health*, *3*(10), e461–e462. https://doi.org/10.1016/s2468-2667(18)30158-0

Clarke, O. (2015, June 19). *Colgate's "80% of dentists recommend" claim under fire | marketinglaw*. Marketinglaw. https://marketinglaw.osborneclarke.com/retailing/colgates-80-of-dentists-recommend-claim-under-fire/

Clayton, T., & Clayton, T. (2021, June 22). *15 Misleading Data Visualization Examples*. Rigorous Themes. https://rigorousthemes.com/blog/misleading-data-visualization-examples/#3_Misleading_pie_chart

Definition of statistics. (n.d.). In *www.dictionary.com*. https://www.dictionary.com/browse/statistics#:~:text=noun,more%20or%20less%20disparate%20elements.

Definition of statistics. (2023). In *Merriam-Webster Dictionary*. https://www.merriam-webster.com/dictionary/statistics

Dickie, G. (2020, November 13). *Why Polls Were Mostly Wrong*. Scientific American. https://www.scientificamerican.com/article/why-polls-were-mostly-wrong/

E. (2022, November 8). *How do you define Data Literacy?* The Data Literacy Project. https://thedataliteracyproject.org/how-do-you-define-data-literacy/

Ehret, T. (2017, June 30). *SEC's advanced data analytics helps detect even the smallest illicit market activity*. U.S. https://www.reuters.com/article/bc-finreg-data-analytics/secs-advanced-data-analytics-helps-detect-even-the-smallest-illicit-market-activity-idUSKBN19L28C

Evans, L. (n.d.). *Adventures In Misleading Graphs*. http://crazytestbl.blogspot.com/2009/08/adventures-in-misleading-graphs.html

Fedak, K. M., Bernal, A., Capshaw, Z. A., & Gross, S. A. (2015). Applying the Bradford Hill criteria in the 21st century: how data integration has changed causal inference in molecular epidemiology. *Emerging Themes in Epidemiology*, *12*(1). https://doi.org/10.1186/s12982-015-0037-4

Federal surveys show no increase in U.S. violent crime rate since the start of the pandemic | Pew Research Center. (2022, October 31). Pew Research Center. https://www.pewresearch.org/fact-tank/2022/10/31/violent-crime-is-a-key-midterm-voting-issue-but-what-does-the-data-say/ft_2022-10-31_violent-crime_02c/

Fluharty, M. E., Taylor, A. E., Grabski, M., & Munafò, M. R. (2017). The Association of Cigarette Smoking With Depression and Anxiety: A Systematic Review. *Nicotine & Tobacco Research*, *19*(1), 3–13. https://doi.org/10.1093/ntr/ntw140

Frey, M. C. (2019). What We Know, Are Still Getting Wrong, and Have Yet to Learn about the Relationships among the SAT, Intelligence, and Achievement. *Journal of Intelligence*, *7*(4), 26. https://doi.org/10.3390/jintelligence7040026

Ingraham, C. (2016, April 11). *The dirty little secret that data journalists aren't telling you.*

Washington Post. https://www.washingtonpost.com/news/wonk/wp/2016/04/11/the-dirty-little-secret-that-data-journalists-arent-telling-you/

Ivermectin shown ineffective in treating COVID-19, according to multi-site study including KU Medical Center. (n.d.). https://www.kumc.edu/about/news/news-archive/jama-ivermectin-study.html

Jeffcoat, Y. (2022, August 24). *How Do Television Ratings Work?* HowStuffWorks. https://entertainment.howstuffworks.com/question433.htm

Kelly, J. (2021, July 25). *Working 9-To-5 Is An Antiquated Relic From The Past And Should Be Stopped Right Now.* Forbes. https://www.forbes.com/sites/jackkelly/2021/07/25/working-9-to-5-is-an-antiquated-relic-from-the-past-and-should-be-stopped-right-now/?sh=485a7ba40de6

L. (2022, January 6). *1.1: What Is Statistical Thinking?* Statistics LibreTexts. https://stats.libretexts.org/Bookshelves/Introductory_Statistics/Book%3A_Statistical_Thinking_for_the_21st_Century_(Poldrack)/01%3A_Introduction/1.01%3A_What_Is_Statistical_Thinking%3F

Lai, S. (2022, June 21). *Data misuse and disinformation: Technology and the 2022 elections*. Brookings. https://www.brookings.edu/blog/techtank/2022/06/21/data-misuse-and-disinformation-technology-and-the-2022-elections/

Lee, S. M. (2018, February 26). *Here's How Cornell Scientist Brian Wansink Turned Shoddy Data Into Viral Studies About How We Eat*. BuzzFeed News. https://www.buzzfeednews.com/article/stephaniemlee/brian-wansink-cornell-p-hacking#.ptrkE1Rxj

LeGare, N. J. P. (2022, March 24). *Link between autism and vaccination debunked*. Mayo Clinic Health System. https://www.mayoclinichealthsystem.org/hometown-health/speaking-of-health/autism-vaccine-link-debunked

Marr, B. (2022, September 28). *The Importance Of Data Literacy And Data Storytelling*. Forbes. https://www.forbes.com/sites/bernardmarr/2022/09/28/the-importance-of-data-literacy-and-data-storytelling/?sh=31cb47ac152f

Maugh, T. H., II. (2019, March 9). *Left-Handers Die Younger, Study Finds - Los Angeles Times*. Los Angeles Times.

McCombes, S. (2023, March 27). *Sampling Methods | Types, Techniques & Examples.* Scribbr. https://www.scribbr.com/methodology/sampling-methods/

McKinney, K. (2014, June 5). *America's favorite foods in 4 charts.* Vox. https://www.vox.com/2014/6/5/5780694/americas-favorite-foods-in-four-charts

Mercer, A., Deane, C., & McGeeney, K. (2020, August 14). *Why 2016 election polls missed their mark.* Pew Research Center. https://www.pewresearch.org/fact-tank/2016/11/09/why-2016-election-polls-missed-their-mark/

Naggie, S., MD. (2022, October 25). *Effect of Ivermectin vs. Placebo on Time to Sustained Recovery in Outpatients With Mild to Moderate COVID-19: A.* https://jamanetwork.com/journals/jama/fullarticle/2797483?resultClick=1

NOVA | The Deadliest Plane Crash | How Risky Is Flying? | PBS. (n.d.). https://www.pbs.org/wgbh/nova/planecrash/risky.html

PolitiFact - Chart shown at Planned Parenthood hearing is misleading and "ethically wrong." (n.d.). @Politifact.

https://www.politifact.com/factchecks/2015/oct/01/jason-chaffetz/chart-shown-planned-parenthood-hearing-misleading-/

Porritt, S. (2023, March 22). *Data Cleaning: Techniques & Best Practices for 2023*. TechnologyAdvice. https://technologyadvice.com/blog/information-technology/data-cleaning/

Roy, A. S. (2021, December 15). *Garbage in, Garbage out: Hidden biases in data. - Aanand Shekhar Roy*. Medium. https://medium.com/@aanandshekharroy/garbage-in-garbage-out-hidden-biases-in-data-e71763b5b79b

Sahagian, G. (2022, March 30). *Analyzing the Used Car Market in 2021 - Geek Culture - Medium*. Medium. https://medium.com/geekculture/analyzing-the-used-car-market-in-2021-27fd460a9067

Scribbr. (n.d.). *The Beginner's Guide to Statistical Analysis | 5 Steps & Examples*. https://www.scribbr.com/category/statistics/

Selvin, Steve (August 1975b). "On the Monty Hall problem (letter to the editor)". *The American Statistician*. **29** (3): 134. JSTOR 2683443

Spurious correlations. (n.d.). https://tylervigen.com/spurious-correlations

Staff, R. (2007, January 17). *Colgate censured over advert.* US https://www.reuters.com/article/uk-britain-colgate/colgate-censured-over-advert-idUKL1654835620070117

Stiglitz, J. E. (2011, March 31). *Of the 1%, by the 1%, for the 1%.* Vanity Fair. https://www.vanityfair.com/news/2011/05/top-one-percent-201105

Stone, C., Trisi, D., Sherman, A., & Beltrán, J. (2020, January 13). *A Guide to Statistics on Historical Trends in Income Inequality.* Center on Budget and Policy Priorities. https://www.cbpp.org/research/poverty-and-inequality/a-guide-to-statistics-on-historical-trends-in-income-inequality

Stratified Random Sample: Definition, Examples - Statistics How To. (2023, March 3). Statistics How To. https://www.statisticshowto.com/probability-and-statistics/sampling-in-statistics/stratified-random-sample/

Taylor, Frederick Winslow (1911), *The Principles of Scientific Management*, New York, NY, USA and London, UK: Harper & Brothers, LCCN 11010339, OCLC 233134

Team, W. (2022, May 20). *Stratified Sampling.* WallStreetMojo. https://www.wallstreetmojo.com/stratified-sampling/

The Hidden Biases in Big Data. (2021, August 27). Harvard Business Review. https://hbr.org/2013/04/the-hidden-biases-in-big-data

The Learning Network. (2020, June 9). *What's Going On in This Graph? | Bus Ridership in Metropolitan Areas.* The New York Times. https://www.nytimes.com/2020/04/02/learning/whats-going-on-in-this-graph-bus-ridership-in-metropolitan-areas.html

The Modal American. (2019, August 18). NPR.

Trufelman, A. (2019, November 11). *On Average - 99% Invisible.* 99% Invisible. https://99percentinvisible.org/episode/on-average/

Ward, P. (2022, August 15). *Frederick Taylor's Principles of Scientific Management Theory.* NanoGlobals. https://nanoglobals.com/glossary/scientific-management-theory-of-frederick-taylor/

What is Bayesian Analysis? | International Society for Bayesian Analysis. (n.d.). https://bayesian.org/what-is-bayesian-analysis/

What's Going On in This Graph? (n.d.). The New York Times. https://www.nytimes.com/column/whats-going-on-in-this-graph

Singal, Jesse *(February 8, 2017).* *"A Popular Diet-Science Lab Has Been Publishing Really Shoddy Research". New York magazine. Retrieved February 20, 2017.*

Endnotes

---

[i] Barnes, B. H. (2013, September 7). *Do left-handed people really die young?* BBC News. https://www.bbc.com/news/magazine-23988352

[ii] Maugh, T. H., II. (2019, March 9). *Left-Handers Die Younger, Study Finds - Los Angeles Times*. Los Angeles Times.

[iii] Barnes, B. H. (2013, September 7). *Do left-handed people really die young?* BBC News. https://www.bbc.com/news/magazine-23988352

[iv] Barnes, B. H. (2013, September 7). *Do left-handed people really die young?* BBC News. https://www.bbc.com/news/magazine-23988352

[v] Baykoucheva, Svetla (2015). *Managing Scientific Information and Research Data*. Waltham, MA: Chandos Publishing. p. 80. ISBN 9780081001950.

[vi] E. (2022, November 8). *How do you define Data Literacy?* The Data Literacy Project. https://thedataliteracyproject.org/how-do-you-define-data-literacy/

[vii] Marr, B. (2022, September 28). *The Importance Of Data Literacy And Data Storytelling.* Forbes. https://www.forbes.com/sites/bernardmarr/2022/09/28/the-importance-of-data-literacy-and-data-storytelling/?sh=31cb47ac152f

[viii] Lai, S. (2022, June 21). *Data misuse and disinformation: Technology and the 2022 elections.* Brookings. https://www.brookings.edu/blog/techtank/2022/06/21/data-misuse-and-disinformation-technology-and-the-2022-elections/

[ix] Marr, B. (2022, September 28). *The Importance Of Data Literacy And Data Storytelling.* Forbes. https://www.forbes.com/sites/bernardmarr/2022/09/28/the-importance-of-data-literacy-and-data-storytelling/?sh=31cb47ac152f

[x] Definition of statistics. (n.d.). In *www.dictionary.com*. https://www.dictionary.com/browse/statistics#:~:text=noun,more%20or%20less%20disparate%20elements.

[xi] Definition of statistics. (2023). In *Merriam-Webster Dictionary*. https://www.merriam-webster.com/dictionary/statistics

[xii] Scribbr. (n.d.). *The Beginner's Guide to Statistical Analysis | 5 Steps & Examples*. https://www.scribbr.com/category/statistics/

[xiii] *Stratified Random Sample: Definition, Examples - Statistics How To*. (2023, March 3). Statistics How To. https://www.statisticshowto.com/probability-and-statistics/sampling-in-statistics/stratified-random-sample/

[xiv] Team, W. (2022, May 20). *Stratified Sampling*. WallStreetMojo. https://www.wallstreetmojo.com/stratified-sampling/

[xv] McCombes, S. (2023, March 27). *Sampling Methods | Types, Techniques & Examples*. Scribbr. https://www.scribbr.com/methodology/sampling-methods/

[xvi] Porritt, S. (2023, March 22). *Data Cleaning: Techniques & Best Practices for 2023*. TechnologyAdvice. https://technologyadvice.com/blog/information-technology/data-cleaning/

[xvii] Beers, B. (2023, March 28). *P-Value: What It Is, How to Calculate It, and Why It Matters.* Investopedia. https://www.investopedia.com/terms/p/p-value.asp

[xviii] *What is Bayesian Analysis? | International Society for Bayesian Analysis.* (n.d.). https://bayesian.org/what-is-bayesian-analysis/

[xix] *Bayes Theorem Application in Everyday Life : Networks Course blog for INFO 2040/CS 2850/Econ 2040/SOC 2090.* (2018, November 19). https://blogs.cornell.edu/info2040/2018/11/19/bayes-theorem-application-in-everyday-life/

[xx] Selvin, Steve (August 1975b). "On the Monty Hall problem (letter to the editor)". *The American Statistician.* **29** (3): 134. JSTOR 2683443

[xxi] *Bayes for days: What to do with signal | Mawer Investment Management Ltd.* (n.d.). https://www.mawer.com/the-art-of-boring/blog/bayes-for-days-what-to-do-with-signal

[xxii] *Federal surveys show no increase in U.S. violent crime rate since the start of the pandemic | Pew Research Center.* (2022,

October 31). Pew Research Center. https://www.pewresearch.org/fact-tank/2022/10/31/violent-crime-is-a-key-midterm-voting-issue-but-what-does-the-data-say/ft_2022-10-31_violent-crime_02c/

[xxiii] *Federal surveys show no increase in U.S. violent crime rate since the start of the pandemic | Pew Research Center.* (2022, October 31). Pew Research Center. https://www.pewresearch.org/fact-tank/2022/10/31/violent-crime-is-a-key-midterm-voting-issue-but-what-does-the-data-say/ft_2022-10-31_violent-crime_02c/

[xxiv] L. (2022, January 6). *1.1: What Is Statistical Thinking?* Statistics LibreTexts. https://stats.libretexts.org/Bookshelves/Introductory_Statistics/Book%3A_Statistical_Thinking_for_the_21st_Century_(Poldrack)/01%3A_Introduction/1.01%3A_What_Is_Statistical_Thinking%3F

[xxv] *NOVA | The Deadliest Plane Crash | How Risky Is Flying? | PBS.* (n.d.). https://www.pbs.org/wgbh/nova/planecrash/risky.html

[xxvi] Frey, M. C. (2019). What We Know, Are Still Getting Wrong, and Have Yet to Learn about the Relationships among the SAT, Intelligence, and Achievement. *Journal of*

Intelligence, 7(4), 26.
https://doi.org/10.3390/jintelligence7040026

[xxvii] *Spurious correlations*. (n.d.).
https://tylervigen.com/spurious-correlations

[xxviii] Calzon, B. (2023, March 1). *Misleading Statistics – Real World Examples For Misuse of Data*. BI Blog | Data Visualization & Analytics Blog | Datapine.
https://www.datapine.com/blog/misleading-statistics-and-data/

[xxix] *PolitiFact - Chart shown at Planned Parenthood hearing is misleading and "ethically wrong."* (n.d.). @Politifact.
https://www.politifact.com/factchecks/2015/oct/01/jason-chaffetz/chart-shown-planned-parenthood-hearing-misleading-/

[xxx] Kelly, J. (2021, July 25). *Working 9-To-5 Is An Antiquated Relic From The Past And Should Be Stopped Right Now*. Forbes.
https://www.forbes.com/sites/jackkelly/2021/07/25/working-9-to-5-is-an-antiquated-relic-from-the-past-and-should-be-stopped-right-now/?sh=485a7ba40de6

[xxxi] Ward, P. (2022, August 15). *Frederick Taylor's Principles of Scientific Management Theory*. NanoGlobals.
https://nanoglobals.com/glossary/scientific-

management-theory-of-frederick-taylor/

[xxxii] Taylor, Frederick Winslow (1911), *The Principles of Scientific Management*, New York, NY, USA and London, UK: Harper & Brothers, LCCN 11010339, OCLC 233134

[xxxiii] Trufelman, A. (2019, November 11). *On Average - 99% Invisible*. 99% Invisible. https://99percentinvisible.org/episode/on-average/

[xxxiv] *Average children per family U.S. 2022 | Statista*. (2022, December 13). Statista. https://www.statista.com/statistics/718084/average-number-of-own-children-per-family/#:~:text=The%20typical%20American%20picture%20of,18%20per%20family%20in%201960.&text=If%20there's%20one%20thing%20the,is%20known%20for%2C%20it's%20diversity.

[xxxv] Trufelman, A. (2019, November 11). *On Average - 99% Invisible*. 99% Invisible. https://99percentinvisible.org/episode/on-average/

[xxxvi] Stiglitz, J. E. (2011, March 31). *Of the 1%, by the 1%, for the 1%*. Vanity Fair. https://www.vanityfair.com/news/2011/05/top-one-percent-201105

[xxxvii] Caporal, J. (2023, February 1). *Are You Well-Paid? Compare Your Salary to the Average U.S. Income.* The Motley Fool. https://www.fool.com/the-ascent/research/average-us-income/

[xxxviii] *The Modal American.* (2019, August 18). NPR.

[xxxix] Jeffcoat, Y. (2022, August 24). *How Do Television Ratings Work?* HowStuffWorks. https://entertainment.howstuffworks.com/question433.htm

[xl] Mercer, A., Deane, C., & McGeeney, K. (2020, August 14). *Why 2016 election polls missed their mark.* Pew Research Center. https://www.pewresearch.org/fact-tank/2016/11/09/why-2016-election-polls-missed-their-mark/

[xli] Dickie, G. (2020, November 13). *Why Polls Were Mostly Wrong.* Scientific American. https://www.scientificamerican.com/article/why-polls-were-mostly-wrong/

[xlii] Mercer, A., Deane, C., & McGeeney, K. (2020, August 14). *Why 2016 election polls missed their mark.* Pew Research Center. https://www.pewresearch.org/fact-tank/2016/11/09/why-2016-election-polls-missed-their-mark/

[xliii] The Learning Network. (2020, June 9). *What's Going On in This Graph? | Bus Ridership in Metropolitan Areas*. The New York Times. https://www.nytimes.com/2020/04/02/learning/whats-going-on-in-this-graph-bus-ridership-in-metropolitan-areas.html

[xliv] Marr, B. (2022, September 28). *The Importance Of Data Literacy And Data Storytelling*. Forbes. https://www.forbes.com/sites/bernardmarr/2022/09/28/the-importance-of-data-literacy-and-data-storytelling/?sh=31cb47ac152f

[xlv] Stone, C., Trisi, D., Sherman, A., & Beltrán, J. (2020, January 13). *A Guide to Statistics on Historical Trends in Income Inequality*. Center on Budget and Policy Priorities. https://www.cbpp.org/research/poverty-and-inequality/a-guide-to-statistics-on-historical-trends-in-income-inequality

[xlvi] The Learning Network. (2018, October 10). *What's Going On in This Graph? | October 10, 2017*. The New York Times. https://www.nytimes.com/2017/10/09/learning/whats-going-on-in-this-graph-oct-10-2017.html?searchResultPosition=3

[xlvii] *2018 and 2019 budgets by technology in*

selected IEA countries and the European Union – Charts – Data & Statistics - IEA. (n.d.). IEA. https://www.iea.org/data-and-statistics/charts/2018-and-2019-budgets-by-technology-in-selected-iea-countries-and-the-european-union-2

[xlviii] Sahagian, G. (2022, March 30). *Analyzing the Used Car Market in 2021 - Geek Culture - Medium*. Medium. https://medium.com/geekculture/analyzing-the-used-car-market-in-2021-27fd460a9067

[xlix] McKinney, K. (2014, June 5). *America's favorite foods in 4 charts*. Vox. https://www.vox.com/2014/6/5/5780694/americas-favorite-foods-in-four-charts

[l] Ehret, T. (2017, June 30). *SEC's advanced data analytics helps detect even the smallest illicit market activity*. U.S. https://www.reuters.com/article/bc-finreg-data-analytics/secs-advanced-data-analytics-helps-detect-even-the-smallest-illicit-market-activity-idUSKBN19L28C

[li] Naggie, S., MD. (2022, October 25). *Effect of Ivermectin vs. Placebo on Time to Sustained Recovery in Outpatients With Mild to Moderate COVID-19: A*. https://jamanetwork.com/journals/jama/fullarticle/2797483?resultClick=1

[lii] *Ivermectin shown ineffective in treating COVID-19, according to multi-site study including KU Medical Center*. (n.d.). https://www.kumc.edu/about/news/news-archive/jama-ivermectin-study.html

[liii] *Spurious correlations*. (n.d.). https://tylervigen.com/spurious-correlations

[liv] LeGare, N. J. P. (2022, March 24). *Link between autism and vaccination debunked*. Mayo Clinic Health System. https://www.mayoclinichealthsystem.org/hometown-health/speaking-of-health/autism-vaccine-link-debunked

[lv] Chiolero, A. (2018). Why causality, and not prediction, should guide obesity prevention policy. *The Lancet. Public Health*, *3*(10), e461–e462. https://doi.org/10.1016/s2468-2667(18)30158-0

[lvi] *The Hidden Biases in Big Data*. (2021, August 27). Harvard Business Review. https://hbr.org/2013/04/the-hidden-biases-in-big-data

[lvii] Roy, A. S. (2021, December 15). *Garbage in, Garbage out: Hidden biases in data. - Aanand Shekhar Roy*. Medium. https://medium.com/@aanandshekharroy/garba

ge-in-garbage-out-hidden-biases-in-data-e71763b5b79b

[lviii] *Biases Archive*. (n.d.). Catalog of Bias. https://catalogofbias.org/biases/

[lix] Fluharty, M. E., Taylor, A. E., Grabski, M., & Munafò, M. R. (2017). The Association of Cigarette Smoking With Depression and Anxiety: A Systematic Review. *Nicotine & Tobacco Research*, *19*(1), 3–13. https://doi.org/10.1093/ntr/ntw140

[lx] Fedak, K. M., Bernal, A., Capshaw, Z. A., & Gross, S. A. (2015). Applying the Bradford Hill criteria in the 21st century: how data integration has changed causal inference in molecular epidemiology. *Emerging Themes in Epidemiology*, *12*(1). https://doi.org/10.1186/s12982-015-0037-4

[lxi] Staff, R. (2007, January 17). *Colgate censured over advert*. US https://www.reuters.com/article/uk-britain-colgate/colgate-censured-over-advert-idUKL1654835620070117

[lxii] Clarke, O. (2015, June 19). *Colgate's "80% of dentists recommend" claim under fire | marketinglaw*. Marketinglaw. https://marketinglaw.osborneclarke.com/retailing/colgates-80-of-dentists-recommend-claim-

under-fire/

[lxiii] Calzon, B. (2023b, March 1). *Misleading Statistics – Real World Examples For Misuse of Data*. BI Blog | Data Visualization & Analytics Blog | Datapine. https://www.datapine.com/blog/misleading-statistics-and-data/#:~:text=In%202006%2C%20The%20Times%2C%20a,visitors%20from%202004%20to%202006.

[lxiv] Evans, L. (n.d.). *Adventures In Misleading Graphs*. http://crazytestbl.blogspot.com/2009/08/adventures-in-misleading-graphs.html

[lxv] Clayton, T., & Clayton, T. (2021, June 22). *15 Misleading Data Visualization Examples*. Rigorous Themes. https://rigorousthemes.com/blog/misleading-data-visualization-examples/#3_Misleading_pie_chart

[lxvi] Singal, Jesse *(February 8, 2017).* "A Popular Diet-Science Lab Has Been Publishing Really Shoddy Research". *New York magazine. Retrieved February 20, 2017.*

[lxvii] Lee, S. M. (2018, February 26). *Here's How Cornell Scientist Brian Wansink Turned Shoddy Data Into Viral Studies About How We*

Eat. BuzzFeed News.
https://www.buzzfeednews.com/article/stephaniemlee/brian-wansink-cornell-p-hacking#.ptrkE1Rxj

[lxviii] Wikipedia contributors. (2023, February 15). *Brian Wansink*. Wikipedia. https://en.wikipedia.org/wiki/Brian_Wansink

[lxix] Aschwanden, C. (2021, March 8). *You Can't Trust What You Read About Nutrition*. FiveThirtyEight. https://fivethirtyeight.com/features/you-cant-trust-what-you-read-about-nutrition/

[lxx] *What's Going On in This Graph?* (n.d.). The New York Times. https://www.nytimes.com/column/whats-going-on-in-this-graph

[lxxi] Ingraham, C. (2016, April 11). *The dirty little secret that data journalists aren't telling you*. Washington Post. https://www.washingtonpost.com/news/wonk/wp/2016/04/11/the-dirty-little-secret-that-data-journalists-arent-telling-you/

Printed in Great Britain
by Amazon